❧ *I live in a glass house*

In the symphony of the biosphere
ecosystems do their riffs
the unseen more powerful than the obvious
we measure and observe;
the creation lives, Frankensystem or Alice in Ecoland
there's a momentum of its own
the people make decisions but hardly call every tune
species invade and conquer new lands
some vanish but their place is soon taken
this biosphere travels on its stomach
it's eat and be eaten
we work to elbow our way in
contemplating how many lobsters, lizards,
goats and people this world can support
we're leaner and meaner,
incredible shrinking biospherians
healthy and hungry
it gives an edge to our lives

— *Mark Nelson*

Life Under Glass
The Inside Story of Biosphere 2

Abigail Alling
and
Mark Nelson
with
Sally Silverstone

Foreword by Joseph P. Allen

 THE BIOSPHERE PRESS™

THE BIOSPHERE PRESS™
A Division of Space Biospheres Ventures
P.O. Box 689, Oracle, AZ 85623 U.S.A.

Executive Editor: Deborah Parrish Snyder
Associate Editors:
    John Allen
    Nicholas Bakalar
    Anthony G. Blake
    Stephen S. Dubin
    Linnea Gentry
    Lynn Ratener

ISBN 1-882428-07-2 (pb)
ISBN 1-882428-10-2 (hc)

Cover design: Kimura-Bingham Design
Typesetting and layout: Kathy Horton
Production: Harrison Shaffer
Printer: Arizona Lithographers, Tucson

 Printed with soy-based inks on recycled and recyclable paper.

Contents

Foreword

Each morning as I leave my home for work, I drive past the outdoor exhibit of a giant Saturn 5 rocket sitting beside the NASA/Johnson Space Center here in Houston. This dinosaur of launch vehicles and its smaller cousins form a kind of Jurassic Park celebrating the time when humans first left our home for bold but fleeting journeys to the moon. In the reddish, early morning glow of the August sun rising with a vengeance above Galveston Bay, the rockets take on a surrealistic look. Did we actually use nine of these behemoths to travel a quarter of a million miles out from Earth?

Such treks are difficult to imagine now that human activity on the space frontier has become so limited. NASA, a federal agency once known for its vision, imaginative engineering, and bold execution, nowadays devotes the bulk of its resources to planning projects which probably won't be implemented. This circumstance should not be surprising, since the demands made by an ever-changing, politically oriented space policy leave the beleaguered agency little extra resources to attempt something new. With such political impediments to overcome, will we ever again undertake space missions out to the moon and beyond?

Dawn breaks over the Biosphere 2 library tower.

And if so, could we establish permanent human habitats out there?

Reflecting upon the heritage of these magnificent Saturn 5 rockets and fretting about the current uncertainties of our space program lead me to think more and more about a totally different kind of spaceship, Biosphere 2.

Strictly speaking, of course, Biosphere 2 is not a spaceship because it does not travel. Nonetheless, I am intrigued by the similarities of experiences of the Biosphere crew members with those of astronauts and cosmonauts aboard their ships in outer space. In spaceship terms, the people aboard this unique creation have been 'underway' for nearly two years now; and one month from the day I write this, they will open the seal which separates them from planet Earth and will exit from their world into an atmosphere as foreign to them as if they had been cruising out to Mars on the far side of the solar system. Does their speed increase now that they are homeward bound? Of course not, since Biosphere 2 has literally not budged one inch since the day the experiment began. Yet figuratively, at reentry minus thirty days and counting, I picture the psychological calendar of the biospherians, having earlier crept forward with agonizing slowness when the Biosphere's food production waned and its oxygen level dropped, now flipping past pages with quickening speed.

I am also intrigued by this enormous experiment because, for the first time, it addresses the missing link of space colonies. By the missing link I mean the understanding of the closed sustainable ecologies needed for human habitation in space. All other technologies needed to live off Earth — rocket travel, for example — were proven during the golden years of the Space

Age. But no understanding about closed ecologies was gained in those years because all space missions to date have relied on a rigid system of consumable stores: food, water, propellants, and so on carried according to a complicated flight plan and meted out piece by piece until they are exhausted about the time, one hopes, of Earth re-entry. Consequently, the single unanswered question still before us would-be space colonists is: can a closed ecological system be devised to be resilient enough to sustain human life for years at a time, yet of a dimension small enough (Earth-sized is clearly too big) to be constructed and maintained by normal human activity?

That this last and perhaps most complicated question would be explored first by a private venture rather than by a well-funded albeit ponderous government research establishment is, in my view, quite remarkable. Regardless of the outcome of these first experiments, the fact that steps toward understanding large biospheres have now been taken is to me both audacious and exciting.

As you read here about life under glass, you may find it difficult to imagine Biosphere 2 as another world. It is, after all, just *there*, separated from our own world only by a simple airlock. To go from one biosphere to another takes just minutes. On the other hand, space is surprisingly close by as well. The space shuttle, for example, travels out beyond the edge of Earth and into the vacuum of space in just over eight minutes, not much different than the time needed to pass through the two hatches and cross the anteroom of the Biosphere's airlock.

But major differences between Biosphere 2 and a spacecraft in orbit do exist — relative size and speed, for example. And there is always the non-trivial matter of the complex physics of rocket propulsion that launch a spacecraft into orbit, inherently dramatic in concept and still bold in execution even in this fourth decade of space travel.

The memory of a rocket launch is not something a person forgets. During powered flight aboard the space shuttle, the engines' roar pervades the crew quarters and the thrust of acceleration holds you against the launch seat at three times your normal weight. After the requisite velocity is achieved, the engines suddenly cut off leaving behind the eerie silence of coasting in unending Earth orbit. The three Gs of acceleration disappear as quickly as the sound. You unbuckle the safety harnesses holding you during launch and float from your seat to the nearest window. You, the space traveler, are now out of this world, privileged beyond all measure to gaze through a window that will forevermore change your perspective of both yourself and your home planet.

Watching Earth from orbit is breathtaking, awe-inspiring, tantalizing, and frightening — all rolled into one complex emotion continually evoked by the panorama before your eyes. Picture yourself floating at that window. Peering out, you watch the oceans and islands and landforms of Earth passing by your window at unimaginable speed. I want to write 'below your window', but in the weightless world of space you have no sense of 'up'. Thus there is no 'above' and no 'below' in orbital flight. You just float at the window and look out on the scene moving past at about five miles per second. Are you speeding by oceans and continents, or are you just hovering in a magical gondola and watching the world turn beside you?

The viewing angle of any part of Earth as seen from the spacecraft window is forever being changed by the relentless pace of orbital mechanics. You are constantly moving your head and hurriedly changing your body position, pressing always closer to the protective glass to catch the last glimpse of your favorite island cluster or to see precisely where your family and friends live there in Houston, marked by Galveston Bay and arrow-straight Interstate 10, the circle of the 610 Loop, and the familiar patterns of the runways of the city's airports. Within hours, the inside of all viewing ports of any spaceship are covered with forehead, cheek, and nose smudges which must repeatedly be wiped away.

Biosphere 2, of course, is not moving. Rather, it is firmly fixed in the hard-packed scrabble of the southwestern desert of North America, a beautiful part of Biosphere 1. The crew within the Biosphere views the outside scene, itself changing with the tempo of the seasons, at a leisurely pace. I suspect that most motion observed outside Biosphere 2 would be the parade of curious Earthlings peering through the glare of the glass walls to catch glimpses of the alien pioneers at work inside their independent ecology.

Along one wall of the living quarters, however, there is a special window through which inhabitants of both worlds can get a closer look at each other and engage in face-to-face conversation of a sort, the words carried by speaker-phones, the images of the people slightly distorted by the internal reflection of the glass. I had the great pleasure of visiting with the biospherians through the conversation window, an occasion which started with greeting them by matching hands, a right hand flat against the left hand and left against right with the palms and fingers separated only by glass. Because of this rite of greeting, the conversation window is constantly covered with smudges of hand prints inside and out quite unlike a spaceship. As far as I know, you will never find hand prints on the outside of a spaceship window.

The book these biospherians have given us here is also a special window of sorts. It gives us insight into the other world in which these unique pioneers have worked and lived for two years.

> — Joseph P. Allen, *former astronaut*
> *Executive Vice-President of*
> *Space Industries International*
> *August 1993, Houston, Texas*

Acknowledgments

Biosphere 2 was the result of a creative and dedicated network of scientists, managers, engineers, and architects who attempted the impossible — and succeeded. But a special word of recognition is due to three people: John Allen, for his vision, boundless energy, and practical savvy; Margret Augustine, for her tenacity in guiding the dream towards its fulfillment; and Edward Bass, for his manifested commitment to the environment and the ecological technologies of the future.

The eight-member crew was only a part of this story; Biosphere 2 needs both a team on the outside as well as the inside. While the following pages give an insight into the challenges of living under glass, the entire biospherian crew is deeply appreciative of the extensive team on the outside whose support made our adventure possible. Without their dedication and vigilance, we could not have completed our task.

CHAPTER 1

The Adventure Begins

We had tested the airlock many times before in trial runs, but this time was for real. At about 8:15 AM on a bright September morning, the eight of us — the first crew of biospherians — stepped into the airlock chamber to begin a journey some of us had been anticipating for seven long years. We had just waved farewell to hundreds of people gathered to see us off before we ducked into this unique compartment. It is about the size of a cargo container, with gray, stainless steel walls and two doors that have portholes like a ship. One door opens out into Earth and the opposite one opens into the Biosphere.

As the metal door swung shut, helping hands tried to push down the large lever on the outside to seal it tight, our last help from outside hands for the next two years. Inside, we pushed down on our side of the lever, but the door wouldn't close. After a few moments Mark Van Thillo called to those outside to step back and then with one decisive swing brought the lever down to seal the door.

A few seconds later, we opened the inner door and entered Biosphere 2. Closure had been accomplished; our self-reliance had begun. The two-year challenge stretched out ahead of us, two years in which we would not go back through that door to

At 8:15 AM on September 26, 1991, eight researchers entered Biosphere 2 to begin its first two-year mission of discovery.

the Arizona desert outside. And no one would come in. What was inside was now all we had.

❧ *Day of closure*

This historic day in our lives, the day of closure, was September 26, 1991. By 6:30 AM everyone — including the eight biospherians, the last-minute work crews, family, friends, and staff — had left the Biosphere and the doors were locked. It was actually at that moment in which the engineer locked the airlock door that our glass world became a separate entity from Earth. No free flow of atmosphere, people, plants or animals, food or supplies would pass between Earth and the inside of the Biosphere again. All the preparations were over. From that moment on, the Biosphere became a distinct material entity, physically closed off from the rest of the universe except for energy and information flows. The physical boundaries were marked by the glass and steel spaceframes above and the stainless steel liner below.

So while the biospherians were taking part in the closing ceremonies in the plaza just outside the airlock, the Biosphere was already on its own. When we headed back towards the airlock at 8:15, we were inserting the final, crucial ingredient into the experiment: the eight humans. Crucial because we were the ones who would either succeed or fail in making this extraordinary laboratory a working reality.

By the time the crew was sealed inside the Biosphere, along with our personal belongings and all the equipment for our mission, the day had already been crammed with activity and emotional intensity. We had woken up at 3:30 AM while the

Biosphere was still dark, long before the sun had risen. We needed the extra hours to prepare for satellite links with the east coast morning TV shows airing three hours ahead of Arizona time. A handful of staff visitors, wanting one last night inside this world they had helped to build, lay curled up in blankets on the couches in the mezzanine. It was the last night that anyone but crew members had been allowed to sleep in.

Those few hours before closure were filled with many 'last' things for us: last walks in the early morning desert air, last hugs from family and friends, last checklists, last treats. A jug of coffee brought by a friend was the last cup for those of us who hadn't already decided to wean ourselves from caffeine. Like many other luxuries, there would be precious little coffee inside the Biosphere; we would have only what beans we could harvest from the handful of young coffee saplings in our orchard and rainforest. And our teas were limited to herbal teas.

Opinions about how to prepare ourselves for such an unusual experiment were decidedly divided. Some crew members, such as Gaie Alling, Jane Poynter, and Mark Van Thillo (known by all of us as Laser), maintained that "the experiment begins when it begins," and they would continue their normal patterns until then. Others, such as Mark Nelson and Linda Leigh (both coffee drinkers), had decided to avoid the bodily shock they had endured each time we started a week-long experimental diet and had gradually cut out caffeine in advance.

The coffee jug exited with the visitors and the final clean-up crew. Along with the jug went the last packaged sugar and Styrofoam cups that we would see for a long time to come. Our very last luxury was a large breakfast of ham, eggs, and bread with butter which we were allowed to enjoy in the peace and quiet of the Biosphere after the ceremonies, final goodbyes, and closure were over. But from then on, all food would be grown, processed, and prepared by our own hands inside the Biosphere under the watchful eyes of Sally Silverstone, our co-captain and also our Agriculture and Food Systems Manager.

☙ Packing our bags

After years of intense concentration on the design and construction of an undertaking as ambitious as Biosphere 2, our own personal preparations seemed insignificant by comparison. So perhaps it's not surprising that some of us waited until the last two weeks before closure to get those needs in order. And many of us weren't even sure exactly what those needs would prove to be. How many pairs of socks, shoes, shorts, pants, shirts, and underwear would we need? What about clothes for special occasions? Would we even *have* special occasions to dress up for? Which books, tapes, cds, photographs, paintings, stereo, TV, mementos, and other personal items should we bring in for our inner nourishment? This was a far cry from packing for a trip to Europe or a summer collecting expedition.

Sally had lived out of a knapsack for years as she worked on various agricultural projects in India, Africa, and Puerto Rico, so material possessions were not a burden she had to deal with. But others had more complicated situations. Roy Walford had cars, a house, and over 1,000 experimental mice in his University of California at Los Angeles pathology laboratory. His cars he loaned to friends, his house he entrusted to his daughter, and his mice became the responsibility of his lab assistants.

Space Biospheres Ventures (SBV), the parent company that created the Biosphere, helped most of us store our clothes, furniture, and other belongings. But Mark Nelson brought all his clothes inside — from fur hats and heavy overcoats to dark suits with black dress shoes to match. Roy brought in his brightly colored *lungis* (one-piece Indian cloth wraparounds for men). Jane brought in a set of wild-colored wigs and masks for parties and other lighthearted moments.

Jane and Gaie shopped for sneakers and blue jeans in the Tucson Mall. Technically speaking, Gaie (whose given name is Abigail) is the Associate Director of Research and Development for SBV and the Manager of Marine Ecosystems inside Biosphere 2. Jane is the Manager of Field Crops and Animal Systems. Gaie holds a bachelor's degree in marine biology and a masters degree in environmental studies from Yale and has tracked whales and dolphins from Greenland to the Indian Ocean to Antarctica. Jane is a gardener, trained in farm management in Australia with a stint of marine ecology on an ocean-going research vessel. Both of them are accustomed to isolation in the outdoors and living out of duffel bags. But even they miscalculated; they wound up with too many blue jeans (twelve pairs in all) but not enough sneakers (six pairs). They had figured that their jeans would be the first things to wear out, because they'd be working daily in the farm area, what we call the Intensive Agriculture Biome (IAB for short). As it turned out, most of us wore out our work shoes before anything else. Jane and Gaie's sneakers all lost their heels and soles and ended up in the scrap box. Indeed, by the end of the two years, bare feet outdoors became a common sight. In the habitat, it had already become a custom, since we had to remove our dirty shoes to avoid tracking mud on the carpeting. Sometimes going barefoot was a pleasure in our tropical world; but it also had the practical purpose of minimizing wear-and-tear on the limited supply of footwear.

Mark Nelson, knowing that half his time would be spent in manual labor in the fields and wilderness, did his shopping in a quick trip to a couple of Tucson thrift shops. A philosophy graduate from Dartmouth College, Mark had spent the last twenty years working on agricultural projects in arid regions of the United States and Australia. He is SBV's Director of Space and Environmental Applications and the Communications Officer for the crew. In the Biosphere he would be helping Linda Leigh, the Manager of Terrestrial Ecosystems, with her work in the terrestrial zones (the rainforest, savannah, and desert biomes). He was also in charge of the waste recycling system and providing fodder for the domestic animals. Piles of new clothes seemed pointless; there would be no one to impress and clearly plenty of ways to get dirty. Laser (Mark Van Thillo) felt much the same way; work clothes would be just fine for him, too.

Jane had the idea to make a special box of new clothes for herself to open on the first anniversary. Mark, although he relied on second-hand clothes, liked the idea and also prepared a box for the one-year mark. This might not be quite like a day's indulgence at the mall, but there would still be an infusion of something new.

We all had our Biosphere 2 jumpsuits, called by some our 'Star Trek suits'. The late Bill Travilla, one of Marilyn Monroe's clothes designers, had been inspired by Biosphere 2 and offered to design something special for us. The result was these lightweight wool jumpsuits in fire-engine red and dark blue. We dutifully wore them for our official closing ceremonies and never wore them again. After losing a good bit of weight from the Biosphere 2 diet, they wouldn't have fit us again anyway.

Although Laser had spent almost no time on his wardrobe, he meticulously prepared a collection of avant-garde music and a library of over three hundred books, rivaled only by Mark's collection. Laser had decided that these two years would be the perfect time to read all the books he'd always told himself he would. His job inside the Biosphere was that of Manager of Technical Systems, responsible for the maintenance and operations of our very extensive infrastructure of machinery. He prides himself on knowing how everything works and being able to fix anything. (He was also co-captain with Sally.) If he had ever managed to finish those five hundred volumes in his apartment, he could have gone on to another thousand in the common library at the top of the sixty-five-foot tower in the middle of the habitat, with a broad range of books in history, art, literature, architecture, ecology and other sciences, and philosophy.

Such was the intensity of the countdown to closure, that some of the special personal items we had planned to bring in never made it. On a late afternoon three days before closure Laser and Gaie had headed for town to stock up on miscellaneous items but never made it. Instead, they pulled over into a gas station to take a nap — the intensity of preparation had exhausted even the inexhaustibly energetic Gaie — and woke up only after the stores had closed. And if anyone still suspects that we stashed bottles of champagne or jars of freeze-dried coffee, we can assure them that we did not! What we could grow in our fields was what we would have, and that principle was never violated.

In addition to our personal wardrobes, we had piles of T-shirts, pants, skirts, and other clothes made of wool which had been donated to us by The Wool Bureau, a company that promotes the use of natural wool products. The Wool Bureau liked the Biosphere 2 concept from early in the project and believed that their products would work well in our tropical world. Their new design actually turned out to be recyclable, biodegradable, and comfortable even in our tropical climate. They also donated our carpets and the brightly colored fabrics on the walls throughout the human habitat area. The walls of the Biosphere 2 Command Room, our semi-circular office and computer/communications hub, are covered by a purple and gray wool fabric; the library walls are azure blue; and each bedroom is a different color of light brown, red, or blue.

Perfume and perfumed soaps or shampoos are not a good idea in the closed, recirculating atmosphere of the Biosphere. First, they would confuse the detailed monitoring which tracks all the small amounts of trace gases that might be in the air. Moreover, in a closed system, compounds from non-biodegradable soaps or shampoos could easily accumulate to a toxic level. To put it bluntly, if we used a product that was not biodegradable, we would be drinking it in our tea within a week. That excluded just about every readily available brand of hand soap, dish soap, detergent, and shampoo.

Sally tested many products, squarely facing the difficult task of satisfying both personal preferences and biological requirements. (She had once managed a hostel for mentally handicapped adults in her native England and sometimes asserted that it was this experience that made her able to survive as co-captain.) Unflappably calm amidst the flare-ups of chaotic activity, Sally had been controller and general manager of the architecture studio for Biosphere 2 during the entire design and construction stage, so settling the soap question was a minor matter for her. The biggest arguments centered around which soap to purchase, and eventually Sally found a way to satisfy everyone. She included a granulated brand of oatmeal soap that dissolved little in the water, as well as an oatmeal and wintergreen soap which melted like soap when wet and could actually produce a lather. She stocked milk crates full of lotion, shampoo, conditioner, wintergreen and spearmint toothpaste, and natural sponges, all of which complied with the need to have no potentially dangerous gas emitters. Those of us who liked fragrances had to collect them ourselves from our herbs in the farm area and plants in the wilderness zones.

Quantities presented yet another problem: how much of the stuff would we need in two years? Some of us, like Linda Leigh and Gaie (who would share one bathroom), ordinarily used up to two containers of shampoo and conditioner (4 ounces each) a week. It seemed unlikely that Roy, who is bald, would need any! Long-haired people would need twice as much shampoo as short-haired Sally or Mark.

There were many such questions to answer before the experiment could begin. What about feminine hygiene? Tampons and pads cannot be recycled, and if four women used these products for two years, the resulting garbage would be impossible to handle. We found an Ohio company called The Keeper that produced a plastic cup which catches the menstrual flow and is reusable. All you do is wash it out. Toothbrushes also aroused controversy. Some dentists recommended twelve toothbrushes; others recommended four to six. In the end, a mix of electric and manual toothbrushes came inside.

Everyone had a thorough dental checkup before coming in, a requirement SBV was very firm about. Our healthcare specialists, Roy Walford, the 69-year-old professor of pathology at UCLA Medical School whose specialty is life extension, and Taber MacCallum, our youngest crewmember and Roy's assistant, had both taken the special U.S. Navy course in emergency dentistry designed for use on ships far from land and without an onboard dentist. Their tales of old-fashioned tooth pulling and other dental tortures had inspired what was probably some of the most conscientious toothbrushing in the history of dental care! Taber was a qualified deep-sea diver and operator of the analytical equipment, but, so far as we know, he had never planned to practice dentistry!

Those with medical problems brought in special equipment. Mark, working on strengthening leg muscles following a knee injury a few years back, split with Roy the cost of a piece of gym equipment for leg presses. Roy also brought a stationary bicycle, a rowing machine, weights, some isometric equipment, and a couple of braces for a neck problem. Mark included knee braces, heating pads, and electric massagers in his kit. At 44, the second oldest member of our crew, Mark found that he was beginning to become far-sighted. So he came in with two sets of reading glasses — one to start with and one for his eye doctor's best guess as to what his eyesight might be after two years.

Since the Biosphere is covered with glass that excludes almost all ultraviolet light, there was no need for sunblock. In fact, we had to take vitamin D pills to make up for the lack of sunlight that the body normally uses to produce its own. We did bring sunglasses, however, because the light can be extremely bright inside; and of course, some used them to look cool.

Another useless item inside is money, although some of it floated in inadvertently in wallets or pockets. There is no money economy inside Biosphere 2. In time, other items came to be used for barter. Eventually, we bartered clothes, time, tools, but not food. Everyone ate their own — food was too precious to trade!

❧ Room for eight

The eight of us were selected from an original group of fourteen biospherian candidates. We're a motley group: five Americans, two Britishers, and a Belgian; four men and four women. Our ages ranged from twenty-seven (Taber) to sixty-seven (Roy) when Mission One began. We are engineers, scientists, gardeners, and explorers.

Although many of us had worked together for several years building the Biosphere, we still came in as eight individuals with unique histories. Gaie Alling, raised on the coasts of Maine and Georgia, is an experienced sailor and expedition chief, the first marine ecologist to swim with a blue whale. Roy Walford, from California, is a world traveler and avid diver in addition to being the well-known author of *The 120 Year Diet* and a professor of pathology at UCLA who had created four transgenic strains of mice and is one of the world's authorities on aging. Linda Leigh, raised in Wisconsin, is a field ecologist as familiar with temperate ecology as she is with tropical, our "wild woman naturalist" whose favorite comic book character is Swamp Thing. Laser Van Thillo is from Antwerp, Belgium, and had trained in industrial mechanics and engineering before heading off to work in India and Central America. Eventually he boarded the research vessel *Heraclitus* first as diver and then as Chief Engineer. Jane Poynter's passion for gardening had led her to range and crop studies in the remote Australian outback, where she began to work with domestic farm animals — unusual interests for someone born to the English upper classes. Mark Nelson, originally from Brooklyn, is a founder of the Institute of Ecotechnics, a small think-tank organization devoted to harmonizing ecology and technology. One of his main concerns for many years has been to bring together space researchers, medical scientists, physical scientists, natural scientists, technologists, and agriculturists from all over the world to work together on addressing environmental problems. Sally Silverstone was the resourceful English co-captain whose background, both academic and practical, in social and agricultural problems of underdeveloped countries probably helped us more than we may have ever predicted when the doors first shut behind us. Taber MacCallum, raised in New Mexico, had also been a world traveler, lived in Japan and Egypt, and been the Chief Diver on the *RV Heraclitus* during expeditions to study coral reefs around Australia and the Red Sea.

Of course, aside from the differences in experience and professional background, we were all different in temperament, a very ordinary mix of 'night people' and 'day people', extroverts and introverts, theorists and practitioners, doers and dreamers all thrown in together to make ourselves into one unified crew with one unified goal. Each of us had our own unexpressed hopes and fears about the commitment we had just undertaken.

Perhaps some of us longed for the scientific joys to be found in this new world or the special intimacy in living closely coupled to the life forms of an intriguingly complex, dynamic system. Some crewmembers had visualized this world as an opportunity for personal transformation — but in what form, who could say?

There were the dark fears, too, which we could scarcely admit to ourselves, let alone the others. Would Biosphere 2 go disastrously wrong? After all, we were stepping into unknown territory. Maybe there was a reason why no one had ever built a biosphere before, practical reasons that we didn't know about. The air could be poisoned by any one of the hundreds of trace gases; dangerous fungi might multiply and invade our bloodstreams. Plagues, locusts, fire, loneliness — we didn't know what really lay ahead.

Although SBV included redundancy and fail-safe backups in the technical designs, we all knew the infallible rule of machines: they break down. If the cooling systems failed, temperatures could rise above 150 degrees under the glass dome in just a few hours, a biospheric oven. And dozens, no hundreds, of other disasters could easily be imagined, including the crew being run ragged working from dawn to dusk to keep everything going. Murphy's Law could undoubtedly apply here as it does everywhere else: if nothing can go wrong, something will; if something goes wrong, everything will. This new world contained hundreds of machines, pumps, motors that could break down, tens of thousands of feet of cable and wiring ready to short-circuit. Were the eight of us heroes or fools for stepping in here?

When we crossed the threshold, we couldn't actually know if we would make it, although we were determined to do so and did not even admit the possibility of failure. From the Russian research in closed systems, the less developed American research, and SBV's own studies, we were confident that it was possible. But not only did we not have all the answers — we didn't even have all of the questions yet! This two-year experiment would be the maiden voyage, the massive shakedown cruise for the most complex ecological experimental apparatus ever devised. If the system worked, Biosphere 2 would provide a powerful new experimental tool for the multi-disciplinary science of biospherics, a controlled microcosm in which to study global ecological processes in detail and as a whole. The unmeasurable would become measurable.

CHAPTER 2

A Day in the Life

✺ *June 2, 1992*

Interviewers and visitors always ask us what a typical day was like inside Biosphere 2. What did you *do* on February 10 or June 2? Did you mark your calendar with any special event that day? Or was it just another Monday or Tuesday?

Even if we didn't check our logbook or diaries, we could say that we spent Tuesday, June 2, 1992 in a way that was unquestionably different from anyone else on Earth. But to us it was a fairly typical day. So here is a look at what filled our hours that day; the explanations of what was going on are brief, because many of these daily tasks were a continuous part of the much larger challenges which the rest of this book is all about.

Dawn broke over the Santa Catalina Mountains at 5:35 AM that morning. No activities were scheduled at that hour, but the crew's early birds were already stirring. Linda had gotten up to do her early morning check of the wilderness areas, including observations on how much had been eaten of the 'monkey chow' put out in bowls for the galagos (the small monkey-like bush-babies) in the lowland rainforest. Sally had made her early morning cup of mint tea and was already on her way to work in the vegetable patch of the agriculture system. Mark, who had also gotten up early, was hand-watering the supplemental crop boxes on the agriculture balcony and cutting some fresh mint

Up with the crack of dawn, the first part of every morning is spent in the fields.

and herbs for the kitchen. He would soon log on to the electronic mail system to pick up the weather report of the last twenty-four hours and record it in his notebook.

By 6:30, Gaie, the breakfast cook of the day, had sounded the official wakeup. She phoned each apartment, allowing the crew a half hour to shower and dress before the hour of work that precedes breakfast.

There's one major difference between the bathrooms you're familiar with and the ones inside Biosphere 2: we have no toilet paper. There's no way that our recycling system could handle the amount of toilet paper that eight people generate in two years. Instead, we use a water squirter that hangs next to the toilet. We found it in a plumbing catalogue designed for Saudi Arabian customers; the Arabs (as well as many other cultures) consider toilet paper far less effective for hygiene and have used water for the purpose for centuries.

Flushing the toilet, of course, doesn't mean that 'it goes somewhere' to be forgotten. All of the water that comes from the human habitat area — from toilets, showers, kitchens, laundries — goes to the basement of the agriculture area to our waste recycling system. Since we check the system every day, we can often tell if a faucet has been left open or a toilet is malfunctioning, because a suspiciously large amount of water will have entered the tanks.

After the wake-up call, Gaie headed for the kitchen to get breakfast going. She pre-heated the oven, boiled water for porridge and tea, and then stopped in at the command room to get the twenty-four-hour report on carbon dioxide to bring to the morning meeting.

By 7 AM, work had started. Sally was milking the four she-goats, and Jane was feeding them. The buck and two bleating kids got their feed, too. Sally fed the chickens their portions of worms and azolla, a high-protein fern that grows on the surface of the water in the rice paddies. Later they would bring the milk, plus eggs collected from the chickens, up to the breakfast table.

Jane checked the status of the irrigation tanks in the agriculture basement and checked the fields to ensure that all the computer-programmed waterings had been triggered successfully. Sally began her daily collection of fresh vegetables, and since it was Tuesday, she also brought up the next week's rations from the basement by elevator — burlap sacks of sweet potatoes, taro, flour, and beans. The supplies and the five-gallon buckets of vegetables Sally had picked soon accumulated in the plaza outside the main double doors leading into the farm.

Linda, Taber, and Mark had taken pruning shears and sickles to the rainforest. Linda and Taber climbed into the spaceframe in the northeast corner to continue cutting back morning glory vines that were shading the trees. Mark was working by the *varzea*, the rainforest stream, cutting and bundling up morning glory vines that crisscross the steep slopes of the banks and wading into the stream to cut and haul out their roots. By the end of the hour, Linda and Taber had brought about twenty-five pounds of the most edible morning glory leaves and vines to the fodder storage bins of the animal bay, our enclosed barnyard area.

Laser logged on to his computer program in the command room to check the technical systems around the Biosphere. A special vibration analysis program gives an early warning of potential breakdowns. He studied the analysis report, completed the weekly maintenance report, then took a quick trip to the basement below the savannah to check on the tanks of water condensed out of the air which passes through the wilderness biomes. Laser is in charge of the rain for the wilderness area (both terrestrial and marine) and must mix the water which drains through each biome (its leachate water) with an appropriate amount of condensed water to make acceptable irrigation water for each area.

Meanwhile, Roy was completing the laboratory workups from the last set of biospherian medical checks. He was also in the process of conducting a stress hormone study, which requires the collection of daily urine samples. He added the fixative agent to the next day's collecting bottle and stored the previous day's samples in the freezers in the genetic and tissue culture laboratory on the mezzanine above the analytical laboratory.

Gaie had by then finished fixing breakfast. Porridge is standard for every breakfast, and that day she had made it with a mix of sorghum and wheat flour, sweetened with ripe bananas and papayas, topped with milk. The rest of the menu depends on the cook's allotments of food. On special holiday and birthday mornings there may be omelets or banana-filled crepes or even a cup of coffee from beans grown on one of the dozen immature coffee trees. (The few beans we grew in the Biosphere were never enough for daily cups of coffee.) This morning, along with the porridge, Gaie would be serving a carrot-cake loaf topped with an icing of milk, banana, and passion fruit; there was also a side dish of beans and sweet potatoes stir fried with chilies.

At 8:00 AM, the kitchen chimes sounded on our two-way radios, and we assembled for breakfast. Our breakfast conversation started off with the usual joking and ordinary remarks, but it also served as a morning staff meeting to update everyone on the progress of the experiment. Sally, as co-captain, called the meeting to order and went over who was on the day's watch and who had cooking duty. She also asked Mark for the weather report, which includes key environmental data: high and low temperatures in all the biomes for the previous day, the relative humidity in the agriculture area, outside temperatures (needed for gauging how to program our air handlers for cooling and heating), outside and inside total light received, and high and low carbon dioxide values at various sensors. Gaie then added the high and mid-point CO_2 for the previous day. There followed a discussion about tactics to deal with CO_2, tactics which had to conform with the SBV research strategy of minimal adjustment of conditions. Should temperatures be lowered in the biomes to lower the soil respiration? What was the status of compost making which releases CO_2? When would the dormant desert and savannah receive their first activating 'monsoonal' rain, which would set off an extra release of CO_2? Sally went over everyone's tasks for the morning agriculture crew, and each crew member outlined his or her day to make sure that all activities were coordinated.

After Sally adjourned the meeting, the watch was officially handed over from Gaie, who was on every Monday, to Mark, the Tuesday watch. Seven biospherians share the watch duties because Laser, as technical manager, has to be on back-up call for all of the others. If anything unusual had happened on Monday, or if there were any alarms during the day, Gaie would have noted them in the logbook. But this day had gone perfectly. Mark took over the watch and checked by radio with the Mission

Control counterpart on the outside, who had also just received the Mission Control watch handed over from the previous day's watch person.

With twenty minutes until the start of the morning work crews, we had some free time to catch up with messages on our computers or the morning news on TV. Gaie did a quick cleanup in the kitchen, loaded the dishwasher, and turned it on. Then she brought a couple of jugs of mint tea to the plaza for morning break.

For all of us the one-hour agriculture crew began at 8:45. Five of us continued on for another two hours. That day we weeded sweet potato, sorghum, and peanut plots in addition to the routine agricultural duties. Laser, in charge of compost making, began his hour by pouring several buckets of animal manure and crop residue into the hammermill which shreds the material into our compost machine and helps speed up the decomposition process. His other responsibility is to feed and water the worm-bed area in the agriculture basement. That morning he brought up a bucket of worms to the animal bay for the chickens. Mark's daily routines include harvesting a bucket of the azolla water fern and cutting fodder for the animals. That morning, he cut elephant grass planted along every available walkway of the agriculture area. He also gathered a bucket of canna that grows in the wastewater lagoons in the south basement and is also used for goat fodder.

While in the south basement where the light spills through a span of glass, Mark checked the wastewater system, which consists of three tanks that receive all the wastewater from our habitat and another set of three for wastewater from the animal bay and the laboratories. When filled, the tanks are closed and bacteria begin the breakdown process. Periodically, by batches, tanks are emptied into the plant lagoons where canna and hyacinth purify the water as they grow. That particular day Mark had to unload part of the lagoon to make room for new waste-water, so before starting the pumps, he checked with Jane to see if the agriculture irrigation tanks were ready to receive the treated water.

Sally continued her round of vegetable harvests, thinning beets from one of our new stairwell planters and picking toma-toes from plants in tubs on the bases of the spaceframe pillars. Linda had been processing wheat from our last grain harvests for the past couple of weeks in the basement. The noise of the thresher prevented her from hearing the radio, so she had told her 'buddy' Gaie to cover any radio calls for her. It's a big world, our three-acre Biosphere, with deep waters, cliffs, and hills, as well as a basement filled with mechanical and electrical equip-ment, so staying in radio contact is important. Having a buddy system helps everyone keep in touch.

We were experimenting with some new varieties of lablab beans that had come to us from a research center in India. Roy was collecting beans from the first plot in which we'd tried them and pruning them back to encourage more flowering. Jane and Taber pruned back the sweet potato plants to stimulate the growth of tubers and collected the fifty pounds of high- protein fodder we needed daily for the goats. The grasses they cut went to the compost machine, and the sweet potato greens went to the animal bay as extra fodder.

Gaie usually spent her first hour tending the orchard, harvesting papayas and figs, pruning citrus and guava trees. Then she checked out the marine systems. This includes looking over all the mechanical systems, recording the ocean and marsh temperatures, and replacing dried algae from the ovens with wet algae newly gathered from the scrubbers, a system which imitates the function of the open ocean to reduce the level of nutrients. She took our little boat 'out to sea' to skim the leaves from the savannah cliff face off the surface of the water. Then she set about cleaning all the skimmate material (a black goo) that had accumulated along the fiberglass pipes of the protein skimmers, another system that removes excess nutrients from the ocean water by bubbling air through pipes.

By a quarter to ten, Laser and Taber were at work on technical maintenance. They cleaned the filters on the basement air handlers that control climate in the savannah and installed new parts to the system that harvests condensed water from the glass over the wilderness areas. Gaie, Jane, Mark, and Sally continued with the peanut harvest. While Mark and Gaie pitchforked the peanut plants into piles, Sally and Jane stripped off the roots and piled the greens into separate buckets. At the end of the crew, the peanuts would go down to the drying ovens, and the peanut greens would be weighed and dried.

By now, Linda had finished threshing and had put the unthreshed wheat sheaves into the oven. After drying, the buckets of wheat grain would go to the seed cleaning machine for the final separation of remaining leaves. Roy spent this second hour in the medical laboratory on the mezzanine floor of the habitat, working on a paper on the oxygen depletion studies he and Taber were carrying out. As work crew ended, the rest of us grabbed a food bucket or sack to carry to the elevator which goes up to the dining room.

Although we are out of physical contact with everyone

except the other seven people inside, parts of our lives are very public. Often a group of visitors watch us through the glass windows. Being caught in undignified positions no longer bothers us. Without self-consciousness we can work barefoot in shorts, even in the most amusing circumstances, such as splashing through the rice paddies splattered with mud while diving for the tilapia fish that live there. We can't hear what people say through the glass unless our ear is right next to it and they're shouting. But sometimes people hold up signs wishing us good luck, and it gives us a boost to see their smiles and thumbs-up gestures. At times when someone wants total privacy, it's not difficult to find seclusion deep in the vegetation of the rainforest or far from the side glass, or to do chores early in the morning or in the evening when visitors are not around.

At 10:45, a break was announced over the radio. Mint tea and roasted peanuts were set out in the plaza. For fifteen minutes, people relaxed on the carpet and cushioned benches or on the first steps of the tower stairway, then dispersed to go on to the next chores.

This was Gaie's day to put on her wet suit and scuba gear for weekly 'weeding' of algae on the coral reef. While underwater, she cleaned the underwater viewing windows and checked the overall health of the reef and fish populations. An on-site security guard watches her through the window, acting as her diving buddy, since we are too few to have her accompanied each time. If anything were to go wrong, the guard would call Laser to join Gaie underwater.

While Gaie checked the coral reef, Sally moved the day's vegetables and a week's supply of staples into the back kitchen to be weighed, logged into notebooks, and stored in refrigerators, grain bins, or freezer. Then she and Jane returned to the agriculture area to examine the crops for insects and disease. Sally had recently released two new types of mite predators, so she also collected a few sweet potato leaves to check on how they were doing. A microscope sits on a counter in the laundry room off the plaza for these examinations. Jane had to spray the rice paddies and grain fields with B.T. (*Bacillus thurigensis*), a bacteria which parasitizes looper worms, from a five-gallon backpack sprayer. This method has successfully kept the worms under control.

By then Linda and Taber, into their second hour of wilderness operations, had moved to the upper savannah to prune back the passion vines and cowpeas that were giving too much shade to the African acacia trees along the savannah stream. In the sand dune area of the desert, Mark took soil cores to measure soil moisture. On the dune sat a squat Plexiglas box nicknamed 'R2D2', a device which continuously measures the carbon dioxide coming out of the soil. We move the machine from biome to biome, to monitor how carbon dioxide efflux changes with the seasons. The desert had had its last rain a few weeks ago, and the soil had been rapidly drying out, with carbon dioxide emissions also dropping considerably. Mark took samples from the first and second feet of soil and weighed them wet before drying them in the agricultural drying ovens. In seven days he would weigh them again to get an accurate reading on soil moisture.

Laser continued his round of preventive maintenance, which today included cleaning a filter in the water systems. He phoned the Energy Center to check on a repair of their backup generator and on our supply of chilled water. The higher the outside temperature, the more cold water is needed to cool down the Biosphere.

We had already cut the dry grasses in the dormant savannah. In a few days, the first rainfall would bring it out of dormancy. The timing of these seasonal climate changes is determined by the SBV research division which then coordinates the Biosphere 2 crew with personnel in Mission Control to program the air handlers.

Between morning break and lunch, Roy re-calibrated sensors for atmosphere temperature and humidity. At 11:30 Sally started a phone linkup with school children in Ohio. Seated in their school library, some sixty eighth graders listened on a speaker phone as Sally talked about Biosphere 2 and then answered questions. Hundreds of schools around the country have used educational materials from Biosphere 2 to learn about world ecology. Many grade-school students have even constructed their own model Biospheres complete with plants and even insects. All eight of us frequently linked up with schools in Arizona and around the country.

By noon, Gaie had showered, changed, and returned to the kitchen to make final preparations for lunch. Like most of the Biosphere 2 cooks, she'd done much of the work the previous afternoon while fixing dinner. With the vegetables cut and the potatoes already baked, she only had to finish her salad. She made a dressing (bananas blended with water and chopped herbs), popped her baked potato casserole into the oven to reheat, and salted the soup of beans, vegetables, and chicken broth which had been slowly cooking in a crockpot since the night before. She then sautéed Swiss chard and beet greens in the wok and mixed up another cold salad of sliced beets and papaya.

Jane and Sally were giving the goats and chickens their mid-day feed. The buck, Buffalo Bill, received a diet lower in protein than the milk goats and had to be locked into a side pen during each of the three daily feedings. The two kids normally kept Buffalo Bill company, but since they were now a week into being weaned, they also were fed a different diet in a separate pen. Buffalo was a veritable Houdini at opening pen doors, so before leaving, Jane double-locked them and tied them with a loop of baling wire.

Lunch time was 12:30. A couple of latecomers let us know by radio that they would be along soon and we prepared plates for them. All our meals are either served on individual plates or put out in servers where it's easy to see each one-eighth portion. *Everything* is eaten, a tribute both to the care with which the cooks prepare the meals and to the appetite we bring to each meal.

Over lunch, we shared more news. We're all especially interested in animal sightings in the wilderness, current news from TV, radio, or E-mail, or what comes over the grapevine from friends or staff whom we've recently been in touch with by phone. Jane brought up the coming arts festival we were planning and who on the outside would be sending in their music, paintings, or poems over the video system. Laser enthusiastically described his encounter with a baby galago in the orchard the previous night; Linda said it was a baby newly born to Topaz. One of our buddies in Mission Control had taken on the weekly task of renting the latest video releases for us. Laser announced that *Lorenzo's Oil* and *Malcolm X* had just been piped in and we all let out a hearty cheer. 'Piping in' a movie means that Laser puts a blank tape into the VCR to record what's being transmitted through the video link with Mission Control.

For an hour and a half after lunch we usually took an informal siesta. After the initial months of adjusting to the new diet, the heavy physical work, and the lower oxygen supply, not many of us actually used the time to take a nap, but relaxed in our rooms reading or phoning friends. It had become just a welcome break in the workday. Everyone in Mission Control knows our schedules, so there are no radio calls unless there is something urgent to attend to. This day Mark's family had come to visit him at the special window we use to meet with outsiders. His mother, who since closure had made the big move from Brooklyn to Tucson, his brother, and his sister-in-law, who live just a few miles away, chatted with him at the window through a speaker phone.

By 2:30, Sally had put the next allotment of food in large plastic tubs in the refrigerator. Taber, the next cook on the rotation, started by checking the blackboard in the back kitchen where Sally had noted the available quantities of some of the staples. These change from week to week depending on our harvests and on Sally's calculations (assisted by computer) of our nutritional needs. Taber's beans have already been soaked overnight, and since the morning they have been slowly cooking in a crock pot. Our foods include many whole foods, so there is more washing, peeling, and soaking than is normally required in typical American meal preparation.

Jane and Gaie put on loose-fitting workclothes for the messy, sweaty job of cleaning two rows of algae scrubbers. Gaie cleaned the Plexiglas containers and wave-buckets, while Jane scraped the old algae off the screens. The growth of algae on the scrubbers is one of the ways we take excess nutrients out of the ocean and marsh waters, but keeping them clean is labor-intensive. The biospherians alternate the job, each doing one to two hours every two weeks. Before they left, Jane and Gaie placed the scraped-off algae in racks in the drying ovens.

Now R2D2 had to be moved to its new location in the lower savannah. First Linda disconnected the long cable which carries its measurements to one of the large computer stations in the basement. Then Mark and Taber coiled the cable and extension cord and gently moved the 'portable' but awkward sixty-pound apparatus over the surprisingly rugged and varied Biosphere 2 terrain. They re-ran the cable and extension cords to the nearest side airvent that drops down to the basement and Linda later rewired the cable to the closest computer cabinet and electrical outlet.

In each biome, a circular plastic ring rests in the ground for R2D2 to sit on. A greased gasket ensures an airtight fit. The instrument has two sensors to measure both the carbon dioxide in the atmosphere and the carbon dioxide diffusing out from the soil. Before R2D2 was constructed by Mission Control staff at the end of our first year, Linda and Taber used to take the measurements manually, taking a syringe of air every six hours over twenty-four-hour periods and then running the air samples on a gas chromatograph in the Biosphere 2 laboratory. Now we occasionally use the manual method to check the accuracy of R2D2 and to take samples of other biomes if R2D2 is occupied elsewhere.

The weekly Mission Control meeting via two-way video hookup started at 3:00 PM. Sally and Laser joined Norberto

Alvarez-Romo, head of Mission Control; Bill Dempster, in charge of engineering systems; and Bernd Zabel, operations manager for the Biospheric Research and Development Center (BRDC). Sally and Laser sat at the V-shaped table in the command room which is outfitted with both the video link and a document reader. They talked over the operational and research activities for the week and a few problems that needed attention from Mission Control. The Mission Control group noted that vegetation was pressing against the spaceframe glass at several points and needed pruning back and that the glass needed cleaning at several places. They made sure that we knew about all upcoming visits by collaborating scientists, guest lecturers, or VIPs in the environmental world.

During the afternoon, we usually work on our weekly and monthly reports or contact people with whom we're doing research projects. Sally finished her weekly report on food production, diet, and nutrition; Jane reported on animal fodder production; and Laser detailed the maintenance program. There are also reviews and projections for the terrestrial wilderness systems (Linda), the marine wilderness systems (Gaie), and the medical and health sensor systems (Roy).

By 3:30, the Biosphere was bustling. Laser was working on technical design problems and the future upgrades during the transition with Ernst Thal-Larsen and Larry Pomatto, Director of Technical Systems, via video. Sally was on the phone with Jim Litsinger, the project's integrated pest management consultant, and Dr. Michael Stanghellini, a pathologist at the University of Arizona who had recently examined roots of our rice, wheat, and sorghum crops to see if nematodes or water-borne bacteria were responsible for the poor yields. They discussed a test proposed

by consultants at the University of Michigan to see if our rice
seedlings were suffering from a more subtle nutrient deficiency.

Mark was on the phone, too, discussing an upcoming
meeting, The Case for Mars Conference, with its organizers in
Boulder, Colorado. Using PictureTel technology, we would be
able to present a 'paper' about our findings and also participate
in a direct discussion with the conference. After he hung up, he
made data entries from the leaf litter and decomposition studies
underway in all the biomes.

Within the hour, Gaie was scheduling the first three
months of the transition phase, the period between Mission One
and Mission Two that would occur between September 26, 1993
and February 26, 1994. She contacted a number of the consultants and collaborating scientists by phone and electronic mail
to arrange for them to come to the Biosphere to make measurements and observations for the completion of their research
projects. Taber spent the afternoon doing routine maintenance
on the automatic sniffer system that analyzes critical gases in the
atmosphere.

It may sound as if everyone was so involved in activity that
there was no room for emotional interaction. But there was —
even if conducting our private relationships was a bit tricky.
We've taken a vow unto death that despite many persistent
questions, we are not going to reveal who was or wasn't involved
in love affairs. Suffice it to say that there were some people in
the Biosphere with such relationships and others without them.
For those without, how is it possible to continue a relationship
with someone on the outside? With ingenuity. Private meetings
at the windows were arranged. While the official biospherian
handshake is two hands matching each other but separated by

the thin three-eighths-inch glass, pairs of lip-prints have been spotted on each side of the meeting window as well. Sometimes they take a while to fade, lingering like the blush of flowers preserved in a diary.

At 4:30, Linda and Mark met in the west arroyo area of the desert to measure plants. This is part of a biomass resurvey they do after the end of each rainy season, to measure how the communities of plants are changing as the desert develops. At the same moment, Sally was discussing with one of the 'food testers' the results of recent trials of recipes from her Biosphere 2 cookbook, *Eating In*. Gaie was deep in conversation with Dr. Jack Corliss, Director of Research for SBV, about the overall research program which includes wide areas of investigations in global modeling, biogeochemical cycles, biomes and ecosystems, systematics, human physiology and nutrition, and engineering.

Meanwhile, a last burst of animal husbandry was going on: by 5:30, Jane was milking goats, Taber had come down from the kitchen to put the buck and kids into their separate pens to feed them, and Sally was taking care of the chickens. After completing the biomass measurements in the desert, Linda connected the cables for R2D2 and checked in again with Mission Control about the carbon dioxide emissions. Mark went to take more soil moisture samples.

By now the official workday was over, but, as usual, a few matters needed attention. Laser was trouble-shooting the electronic sensing devices of the fire alarm system which had been giving false alarms due to high humidity. Gaie and Mark had to review a press release written by the public affairs office. Sally returned a call from a Tucson journalist requesting her reactions to a National Academy of Sciences warning of the danger of pesticide levels in fruits and vegetables, especially to children. She used the opportunity to discuss the relevance of our agricultural system, which doesn't rely on chemical pesticides and is nevertheless extremely productive. The journalist also wanted to come out for a photograph of Sally with some of her new beneficial insects, so she arranged a time for the following morning.

Roy was cook's helper for the day and had begun helping Taber by washing pots, making tea, and setting the table in the frequently frantic last minutes before meals are served. Gaie and Mark, hungrily anticipating dinner, took the opportunity to work on this book, which they had been writing since the early days of the experiment.

Taber and Roy served dinner at 6:45 PM. Taber had made one of his characteristically beautiful meals, featuring a flour shell into which he had put a delicious chili with lablab beans, chives, taro, and green banana. There were sliced beets, a mixed vegetable dish, baked sweet potatoes with a banana sauce, and tossed salad. Dessert was sweet potato pie topped with slices of fresh fig.

Less than an hour later, Taber and Roy were doing the kitchen cleanup, which included sweeping and mopping the floor. They had cooked the day's food scraps in a large pot on the electric stove, and the leftovers were now ready to go down for the goats. Taber started on tomorrow's breakfast. He began slow-cooking the porridge overnight in a crock pot and then put rolls and sweet potato patties on baking sheets in the refrigerator along with other foods for tomorrow's lunch.

Sally and Laser worked together in the command room

summarizing the day's events in the captain's log and sent it via the network to Mission Control and SBV management. At 8:15, Mark started his night rounds — a twenty-to-thirty-minute tour through the agriculture, rainforest, savannah, and desert, noting the manual thermometer readings in each area. He also turned off some lights that had been left on and then checked the algae scrubber room, the recirculating fans, and the wave machine in the ocean. He also glanced at the alarm screen in the savannah tunnel for any temperature red alerts and then checked another alarm board nearby for any yellow lights indicating technical malfunctions. He then copied his readings in a log book kept near the alarm screen monitor.

Linda had settled in with her computer to log onto the WELL, an electronic communications network run by the Whole Earth Catalogue. She checked over a general bulletin board for useful information and examined WELL's array of electronic conferences. Linda frequently contributes to conferences on Biosphere 2, as well as many others on politics, the environment, and cutting edge technologies. Roy was on the phone to friends in Los Angeles. Taber and Jane were watching television, and Gaie was reading.

Our apartments are our castles. When we want to be alone, this is where we retreat. Each biospherian has a two-story apartment with a downstairs area that includes desk, bookshelves, TV/VCR/ and radio/CD/tape player, sofa and easy chair, and clothes closet. Up a circular staircase is the loft bedroom with clothes bureau. There are ten apartments in Biosphere 2, the extra two for visiting scientists who may use them in the future.

Before closure, the crew had the opportunity to decorate their 'pads' with whatever they wanted. So we had our favorite paintings, souvenirs from travels around the world, books, and music collections. We also chose rooms with views that we liked — though all the views are out of this world. Some look out over the agriculture area and south to the Santa Catalina Mountains. Others look out over the ridge to the west and are good for sunset-watchers. Some have a view of the area in front of the habitat, where gleaming, white tents spring up during special events at Biosphere 2.

Laser had transformed his apartment into a video studio with gear for making documentary films. A piece of blue plastic in his window mystified visitors until he explained that he was protecting his equipment from harsh sunlight. Linda decked out her apartment with beautiful artifacts from native cultures around the world. Roy's apartment was decorated with art pieces from artist friends. Mark's was dominated by his huge library of books, Australian aboriginal carvings, and colorful paintings. Bathrooms are shared between two apartments.

Most of the crew kept a personal record of their experiences, and before turning in at about 9:15, Mark made his daily entry in his computer. Others kept long-hand notebooks. By 10:00 PM, the hallways of the habitat were dark, although light seeped out under the doors of a few apartments.

For all eight of us, another day of the 731 days we would spend in Biosphere 2 had ended. As we drifted into sleep, our world continued to hum around, above, and beneath us. Almost all of the crew have remarked upon the special intensity of dreams inside Biosphere 2, but no one noted any particularly unusual dream that night. A few of us may have noticed the sliver of a new moon rising outside, refracting through the beautiful geometric glass sky of our miniature world.

CHAPTER 3

Monitoring the Environment

❧ Our own EPA

If the city seems smoggy, maybe you can drive out of it to the country. If it's smoky near a factory, you don't go near it unless you live or work nearby. Oil spills and nuclear plant leaks—well, they're nasty, but maybe they're far away and there are other things to worry about. But in Biosphere 2, this way of thinking will never do. In a small living system, it is vividly clear that everything we do has immediate and potentially devastating effects on our surroundings. So we must police the environment, ever alert for anything that could pose a health threat to ourselves or to any of the other life forms. That means paying especially close attention to the quality of air and water, the first of the critical life-support elements to reveal the onset of pollution.

Part of the pre-closure preparations involved the formulation of 'mission rules'. These resemble the rules that would be outlined by an environmental protection organization. Among the guidelines were air and water temperature limits, the allowable concentrations of potentially toxic molecules or heavy metals in the drinking water or air, and the management of nutrient cycling. There were many unknowns, and many more partly knowns, in this process. After all, the guidelines can only be set by measuring the natural world in minute detail, the

Mark Van Thillo calibrates a humidity sensor in the savannah.

means of which have only just begun to be developed during the past few decades. In the case of specific air and water contaminants, the SBV research adopted the standard warning levels used by OSHA (Occupational Safety and Health Administration) for industry, by EPA (Environmental Protection Agency) for the environment, and by NASA (National Aeronautics and Space Administration) for astronauts. One of the long-term purposes of the Biosphere 2 experiment is to help develop better environmental guidelines for use everywhere.

Management of our closed system requires very precise analyses of our environment. We depend on getting those measurements quickly enough to make the right decisions by a combination of electronic sensors, automated analytical facilities, and manual laboratory procedures. The Biosphere was originally equipped with a state-of-the-art analytics laboratory capable of making the precise measurements of small concentrations of compounds, designed to run without relying on outside support. All other existing laboratory systems use large quantities of toxic chemicals that we just couldn't release safely into our system. So working with the experts, SBV had coordinated the design of a lab that uses no toxic chemicals and recycles any by-products in a waste recycling system similar to the one that handles the wastewater from the human habitat. In this way we've been able to avoid the paradox of most environmental laboratories that, even as they help identify toxic pollution problems, they create them themselves.

After the first ten months of closure, the Scientific Advisory Committee recommended that we export research samples for analysis in SBV's outside lab and other independent laboratories. The reason being that we could increase the amount of data we were getting from the experiment and also decrease the time we had to spend doing the analyses inside. Fourteen months into the experiment, SBV made the decision to export much of our analytical equipment and set it up in our Biospheric Research and Development facilities outside the Biosphere. So even though the merit of the new laboratory system was proven, in the end it proved more efficient to have this superb equipment operated full-time on the outside.

The wonders of modern analytical technology allow most air and water compounds to be detected even in trace amounts. Some can even be detected when they are as low as a few parts per billion or even parts per trillion! This allows a safety margin in detecting increases of toxic compounds long before they approach threatening levels. These advanced systems also enable us to do some fascinating detective work.

A few months after closure we experienced a case in point. The research department noticed that trace gases which commonly come from the glue used to seal PVC pipes were persisting in our atmosphere. Yet, from our earlier experiments, we knew that these gases should have been gradually cleaned up by natural ecological processes. This meant that there was something inside the structure that was still releasing the gases. We immediately searched the Biosphere for any containers of glue, primer, or sealant that may have been forgotten by the construction crews who put the pipes together. To our amazement, we found four separate containers of PVC glues and solvents, in remote corners, which were slowly leaking fumes into the air. Once these cans were sealed off in containers and removed to the west air lock, the levels of the trace gases we had been tracking in the atmosphere finally declined.

Because there are so few sources for toxins in our atmosphere (we have no gasoline, smoke, paints, or even perfumes), it was primarily welding and technical repair operations involving PVC solvents or glues that caused small spikes of trace gases to appear. During the first year, the SBV research and analytical staff debated whether the atmosphere and ecological systems, which ultimately metabolize the gases, could support necessary repairs to some of our mechanical systems. The discussions pushed us to devise strategies to minimize the impact on the atmosphere.

In fact, we had to think about these things all the time, about everything we did. The consequences of seemingly innocuous actions like opening a can of PVC glue are immediate and real. There are no 'anonymous' actions. Even if someone didn't tell us that they did some midnight painting or gluing, we would know by observing which trace gases had risen in our air since the last analysis.

⌘ Sniffing and sipping

In order to measure some of the gases contained in our atmosphere, SBV set up an elaborate device called the 'sniffer system'. This is an automated system located in our analytical lab in the habitat with sensing stations located throughout the Biosphere. These continuously monitor the gases that could present health concerns as well as those which we thought would show significant changes during the experiment: CO_2 (carbon dioxide), O_2 (oxygen), NO_2 (nitrous oxide), NO_x (other nitrous gases), SO_2 (sulfur dioxide), CH_4 (methane), O_3 (ozone), and H_2S (hydrogen sulfide). Each gas is detected by an instrument that 'sniffs' the air every fifteen minutes. The research staff can call current values from the data base or generate graphs to show trends over time. To measure other trace gases, we collected samples of the atmosphere every two weeks and analyzed them with the equipment in our lab or sent out to collaborating labs.

To lessen any potential problems to human health from the gases, the Biosphere 2 design team had chosen to use natural building materials such as wool and wood. We also tried to avoid materials known to have nasty trace-gas emissions. Although we occasionally had to use glues, sealants, or grease to repair piping or maintain pumps, we had to make an effort to keep their use to a minimum. Before this experiment, however, no one could fully predict the magnitude of the problems we would face. Until now such a complex system of technology and ecology had never before been sealed in an air-tight facility. For example, we wondered what methane would do because our world includes marshes, rice paddies, compost heaps, and even our own digestive systems — all methane emitters. Our microbial ecologists predicted that the methane-eating bacteria in the soils would increase to keep the methane in balance. And indeed, after an initial period of increase, levels of methane stopped rising and remained safely low and steady throughout our two-year closure. One trace gas that has slowly but steadily increased since closure is nitrous oxide, although it is still at safe levels. This is the subject of much humor among us as nitrous oxide is also known as 'laughing gas'. But though we have about thirty times as much nitrous oxide as Earth's environment, it's still measured in tens of parts per million. To get the full dentist's chair effect, it would have to be well over half of our air.

But what about the smells? Even minute amounts of a gas

can create a stink or a fragrance. Fortunately, we had the opportunity to verify that we had not just become accustomed to our air, thinking it smelled fresh when, in fact, it really didn't. During our import and export exchanges of research materials, those on the outside waited eagerly to come into the airlock chamber to take deep breaths and catch a whiff of our atmosphere. They reported that the smells were delightful, similar to the deep organic smells of rich farmland or a rainforest.

The variety of natural smells was one of the many delightful surprises we encountered when we conducted our first Test Module experiments. (The Test Module was a small 'prototype' where we did some of our first experiments before building Biosphere 2.) People who lived inside the structure for days or weeks always describe upon the rich natural smells as like a freshly cut field of hay or a rich, earthy aroma. Even more intriguing for all the biospherian trainees and SBV staff who spent a minimum of twenty-four hours inside the module was how smells differed depending on the time of day.

A variety of aromas was deliberately included in the design of Biosphere 2. All the biomes have different smells as well as different moods, light, and humidity. Many of the plants in the desert were chosen specifically for their pungent aromas. These aromas may either be part of the plants' chemical defense systems to discourage grazers from munching precious leaves or part of its attracting system, to lure pollinators. For us, they made a trip to the desert a sensory feast. It was a delight to brush against the plants and release the aromas. In the smell-conscious environment of the Biosphere, other crew members could always guess where you'd been when you came back to the dinner table.

For monitoring critical water parameters, we designed an automated 'sipper system' similar to our sniffer system for air. It's located in the technical basement below the savannah where it receives water sequentially from the ocean and marsh and analyzes the concentration of nitrates and nitrites. These nutrient measurements are critical for the management of coral reef and marsh health.

Coral reefs are often referred to as the 'rainforests of the ocean' because of their dazzling diversity of coral and fish life. But curiously enough they thrive in extremely nutrient-poor waters and are quite sensitive to changes in nutrient content. An interesting example of this occurred during the summer monsoons in 1990 when the ocean was being stocked with lagoon grasses from Florida and reef rock turf from the Caribbean. The monsoon season in the Sonoran Desert brings repeated heavy rain storms. Since this section of the Biosphere was not yet sealed overhead, the rains often dumped between 6,000 and 12,000 gallons of water into the fledgling ocean at a time. Invariably the water would turn a pea-green color the next day, an indication of an overabundance of plankton. We found that the water was quite laden with nutrients, with many times more nitrate in it than would be found in reef water! Even though the city of Tucson is about thirty miles away, nitrate-laden smog had contaminated the rain as it fell.

The relative acidity or alkalinity of the ocean is described by a pH number. Initially, the ocean pH was targeted to maintain levels within the range generally found throughout the world's reefs, from 8.2 to 8.4. But the ocean could not maintain this level because of the higher, frequently changing levels of carbon dioxide in our atmosphere. The ocean water absorbs carbon dioxide, which lowers the pH of the ocean. So over time, we had to lower the alarm thresholds, since we were routinely operating at pH levels lower than those in natural conditions. We were relieved to see that the coral reef community tolerated the changes. In fact, much of what we are learning centers on re-defining ecological tolerances as we discover how the biomes react to conditions not naturally found in the environment. By recreating complex biomes isolated in Biosphere 2 and exposing them to new environmental conditions, we are actually experimenting with ecology. *That* to us is the really exciting stuff here.

Our drinking water is condensed out of the atmosphere, either collected from the water that condenses on the glass or caught in trays in the air handling units. Depending on the season, we get from 4,000 to 8,000 gallons of pure water each day. This is possible because water moves quite rapidly through our atmosphere by means of the transpiration and evaporation cycle. Transpiration is a process by which plants pass enormous quantities of water through their leaves. For example, a corn plant will transpire an estimated 100 gallons of water for each gallon that it actually uses for growth and fruiting. In addition, the large surface area of the ocean and marsh allows a considerable amount of water to evaporate, which also increases the humidity. On days when the humidity is very low we estimate that approximately 2,000 of the ocean's 900,000 gallons of water can evaporate. Over the course of a year almost all of our ocean will evaporate and be replaced by other Biosphere 2 water! During this process most salts or other trace compounds are left behind, so this condensate water is quite pure and suitable for drinking.

Once the condensate is collected, we can pass it through an

ultraviolet/hydrogen peroxide system to kill bacteria. We chose this kind of sterilizer rather than one using chemical additives, because our drinking water eventually winds up in our waste-water lagoons and the irrigation supply. Even small amounts of chemical purifiers, such as iodine or silver, would build up to dangerous levels over time. On the scale of a century, the planned lifetime for Biosphere 2, this becomes an unacceptable option for it would eventually prove disastrous. Our internal EPA rules take this longer view very seriously, as we don't want to pass on any environmental time-bombs to future crews.

All of these safeguards put together have given us a safe, continuously viable system. We conduct a full analysis of our drinking water twice a month, while the wastewaters from soils, wilderness condensate, streams, and the rice paddies are tested once a month.

Overcoming the sick-building syndrome

Air is complex. Although it mostly consists of nitrogen and oxygen, the small concentrations of other gases it contains are very significant for the living systems that breathe it. The primary gas components of air are nitrogen (78.08%), oxygen (20.9%), argon (0.93%), and carbon dioxide (0.035%), but there are significant though small amounts of what are called trace gases, significant because some may become toxic if their concentration rises even slightly.

You don't have to investigate the Biosphere 2 experiment too deeply before it becomes obvious that maintaining a healthy atmosphere is our most important task. Even with our high roofs, there's a relatively small amount of air in our world compared to the mass of soils, plants, and man-made materials. Our experiments in the Test Module showed that even rigorously evaluated materials still produced some outgassing into the atmosphere. This outgassing can produce a bewildering mix of chemicals. Even in spacecraft (such as Skylab or the Space Shuttle) where materials are chosen to minimize outgassing, hundreds of trace compounds have been detected in the cabin air, just as we found in the Test Module.

Trace gas build-up has always been one of the most serious potential problems in small life-support systems. Before closure, Mark spoke with a space scientist who told him that when the Orbiter space shuttle's doors were first opened at Edwards Air Force Base after a week in space, a member of the ground team threw up because the smell inside the space craft was overpowering. The obnoxious smell came from the notorious zero-gravity toilet which then combined with other odors accumulating in the tightly sealed cabin. In the Russian Bios-3 experiment, even the plants stopped growing due to the toxic buildup.

These types of problems aren't confined to spacecraft — many ordinary homes and office buildings exhibit the 'sick-building syndrome'. People quickly get used to bad odors — so much so that we often have to be informed of a smell by someone with a 'fresh nose' who has just entered the environment. It's not just that the smell is unpleasant, it can also be unhealthy. The interrelationships of plants and soil microbes are beneficial in that they are uniquely able to clean up these trace gases. More wide-spread use of indoor landscapes in offices as well as of potted house plants would go a long way towards solving the sick-building problem.

To ensure that we would not be plagued with these problems in Biosphere 2, we developed and patented a system based on an engineering technique first developed in the early 1900s. A discovery made in Europe showed that pumping the nasty odors from food-processing factories through a nearby field of soil purified the air. They called this a 'soil-bed reactor'. SBV turned this discovery into an ecological apparatus that can purify the air of any toxic elements through the form of a small commercial unit, called an Airtron™ that cleans the air using the combined activity of houseplants and soils. This soil-based system, in addition to allowing us to recycle wastes through composting techniques, also allowed us to control toxic gases in the Biosphere 2 atmosphere.

At first, a small portion of the soils in all the biomes were designed as soil-bed-reactor back-ups. But finally, to be able to deal with any eventuality, the entire agricultural area was constructed with an Airtron™ system to allow air to be pumped through all the soil. If necessary, we can force the entire Biosphere 2 atmosphere through the farm soil in about a day. Part of the beauty of the system is that whatever trace gases exist in the air serve as food for microbes and these microbial populations will increase until they have brought down the concentration of trace gases. The increased air flow can even help minimize pockets of waterlogged or anaerobic soil. It's a marvelous technology born of billions of years of microbial evolution!

To date, the natural systems in Biosphere 2 have functioned so well that so far we haven't needed to use the soil-bed reactor inside Biosphere 2. We activated it only for test purposes. That isn't to say that trace gases are not present, because of course they are. But the Biosphere's ecosystems have been able to maintain them at very low levels posing no health concern.

✲ *The tiger of Biosphere 2*

Because of the much smaller ratio of atmosphere to soil in Biosphere 2 compared to Biosphere 1 (Earth), the design had to take into account that soil respiration would, at least in the first few years, increase CO_2 levels to as high as 4,000 parts per million during winter months. In addition, the smaller atmosphere would mean a far greater daily surge (about 600 parts per million) of CO_2 than the one or two parts per million in Earth's atmosphere. So we paid very close attention to carbon dioxide.

On a normal day of sunshine CO_2 can drop by as much as 600 to 800 parts per million and then rise as much during the night. By comparison, global CO_2 only varies by ten to fifteen parts per million over the course of a year. But just as the citizens of Earth are concerned about the slow and steady increase in CO_2 caused by industrial emissions and the loss of forests (the greenhouse effect), the Biosphere 2 EPA also has CO_2 near the top of its list of concerns. It could cause us problems either by getting so high (over 1 percent or 10,000 parts per million) that it causes animal and plant toxicity, or by dropping so low (anything much below 250 parts per million) that plants would be limited in their ability to make new growth through photosynthesis. Biosphere 2 has operated between 1,200 and 4,000 parts per million as its daily average depending on the time of year. If the days were continuously sunny, it appears that Biosphere 2 CO_2 would stabilize at around 1,800-2,000 parts per million or slightly lower.

Every fifteen seconds, a new value for CO_2 from continuous sensors distributed throughout Biosphere 2 (agriculture, habitat, rainforest, savannah, and desert) appears on the 'global monitor' computer screen. Biosphere 2 is so sensitive that simply by looking at the numbers changing on the screen we can tell whether a cloud has momentarily blocked the sunlight or a brief patch of sunshine has opened up on an otherwise overcast day.

Carbon dioxide is on a roller coaster in our small world because of its sharp changes from day to night and the way it can quickly rise over successive days of cloudy weather. Helping the system manage the CO_2 levels was one of our major challenges. It became known as 'the tiger of Biosphere 2'. And as the saying goes, once you catch a tiger by its tail, never, never let it go. So in addition to close monitoring of its levels, the research department developed a number of strategies to boost photosynthesis and minimize respiration.

Reducing respiration, which releases CO_2, also helps correct a 'greenhouse effect'. The best way we can help to lower CO_2 is to maximize plant production so that the plants fully utilize the precious 'sunfall' that enters our world. For example, trees are pruned and vines removed from the glass to ensure that sunlight can reach the understory vegetation. This stimulates new growth that takes up CO_2 more efficiently than does old growth. Biomass is cut from fast-growing areas of the wilderness biomes such as the ginger belt perimeter of the rainforest and the tropical grasses of the savannah. The storage of the pruned vegetation keeps the carbon it contains from being respired back into the atmosphere as CO_2. We also stop composting in the agriculture during winter since this process also releases CO_2.

In the sunny spring season of long days, we resume composting and return the stored plant material to the ecosystems as mulch.

To deal with the worst CO_2 rise at the winter solstice period, we built a unique recycling system which mimics Earth's natural geological processes. The recycling system is operated from time to time in the late fall and early winter to pull CO_2 out of the air and through a series of simple chemical reactions, changing it into a powdery limestone, calcium carbonate. When CO_2 is needed back in the atmosphere, this limestone can be heated in a furnace to re-release the CO_2. The benefit of the system is that you recycle the starting chemicals and begin all over again. This is analogous to what occurs on Earth over centuries: CO_2-storing limestone forms on the ocean floor and $CO2$ is released from the Earth's interior in volcanic emissions.

The unprecedented cloudy winters during both years of Mission One proved to be as good a test as any of our ability to track and manage CO_2. In 1991 and 1992 the El Niño ocean currents caused the southwest United States to receive an unusual number of fall and winter storms. In January and February of our second year, southern Arizona suffered record floods from a series of successive winter storms. Despite all the cloud cover, our maximum monthly CO_2 average never got very far above 2,700 parts per million the first year and 4,000 parts per million during the second year. When the sun finally returned in March of 1993 for our final six months, it was clear that Biosphere 2 had managed to handle the problem. The research department anticipates that the Biosphere itself will more completely self-regulate the CO_2 in future years as the young soils mature and have less organic material to produce CO_2 during oxidation. Also as the trees grow larger, they will gain a greater capacity for photosynthesis.

Oxygen was also monitored closely, using two instruments in the analytical lab. One was a Teledyne oxygen analyzer, a standard piece of equipment for determining safe oxygen levels. The early readings indicated oxygen levels in the Biosphere were more than twenty percent. But because this instrument was designed as a rough indicator of the rise or fall of oxygen levels, Taber was scheduled to make a routine calibration check every six months using the more sophisticated and accurate gas chromatograph.

In February 1992, five months after closure, Taber was shocked to see during the calibration check that the oxygen levels in the Biosphere were falling markedly. It became clear that the continuous analyzer for oxygen had been giving inaccurate readings since closure. This drop in oxygen was particularly startling because during all the Test Module experiments we had never experienced a fall in oxygen levels. Nor had it been reported by other life-support facilities, so it had not been an area anyone in the field considered critical. But the oxygen was indeed declining and it rapidly became our number one concern. John Allen, SBV's vice-president of Biospheric Development, and Bill Dempster met with Wally Brocker, geochemist at Lamont-Dougherty Institute, and began a research program under Dr. Brocker's guidance with a young doctoral student, Jeff Severinghaus. Several other scientists specializing in geochemistry and physiology joined in. It was decided, with the consent of the biospherians, to 'ride the oxygen down' because of the extraordinary research opportunity. The oxygen level would be raised the moment any health concerns arose.

Finally, the time for a difficult decision came. On January 13, 1993, after several members of the crew began experiencing physical difficulties, some 31,000 pounds of liquid oxygen were injected into the air over a nineteen-day period to boost the levels of oxygen from approximately 14.2% up to 19%. We didn't want to add outside air because the mix of gases would disturb our other research on the composition of our atmosphere. We needed pure, measurable oxygen in the form of liquid oxygen, the same form of oxygen used to fill hospital oxygen containers. A flotilla of trucks carrying liquid oxygen came out from Tucson; the liquid oxygen was converted to gas by vaporizing the liquid with a heat exchanger before passing it through a connection in the west lung to then spread throughout the entire sealed structure.

After examining his medical data, Roy recommended that we add to the mission rules a lower limit for oxygen of sixteen percent. Above this level only a few biospherians had experienced early indications of high altitude sickness — namely, difficulty sleeping at night without extra oxygen being pumped into their rooms. For these people, extra oxygen was provided by an apparatus in the Biosphere 2 lab. Later, we again injected liquid oxygen one more time so that scientists and technicians unaccustomed to our lower oxygen levels could go in and out of the Biosphere during the transition period between our mission and the next.

✺ *Our endangered species act*

During these challenges of carbon dioxide and oxygen, we also had other issues engaging our attention from the beginning. It was highly uncertain whether we could succeed in maintaining distinct biomes over time and whether our initial richness of species could be maintained. Some people warned us that Biosphere 2 might decline into a soup of algae as the higher plants and animals died off. Some leading ecologists even predicted a severe loss of species. About a year before closure the two great founders of systems ecology, Eugene Odum and his brother Howard T. Odum (H.T.) visited Biosphere 2. H.T. laid a wager that within the first two years we would lose eighty percent of our initial species, and that the remaining twenty percent would ignore our carefully crafted design for different biomes and organize themselves into something completely different. He didn't doubt that Biosphere 2 would work as a total system, but it would be nothing like the world we had designed. Linda bet that losses would be less than twenty percent after two years.

Dr. Odum visited again just before our first-year anniversary. With relief and great good humor, he conceded that he had lost the bet. He was impressed that we were evidently holding on to sizable biological diversity and surprised to see that the biomes still showed very distinct and different characteristics. One factor he may not have fully appreciated is the sophistication of the environmental technologies that assist the biomes in maintaining proper temperature ranges, generate waves and currents, and produce rainfall. All of these factors can be adjusted to fit the needs of the ecosystems.

As early as 1966, the Odums had unsuccessfully urged NASA to take a more ecological approach to designing life-support systems. H.T. has long advocated taking a sealed greenhouse and literally tossing in plants, animals, soils, seed, and people as an experiment to demonstrate the power of ecological communities to organize themselves. In such an arrangement, there would probably be sizable extinctions before the system settled down. Our strategy actually did include a 'species-packing' approach; we included nearly 4,000 species of plants and animals and uncounted microbial species. We did it this way mainly because no one knew which species would be lost or which would do well. The ecological designers tried to include a number of species in each biome that performed the same ecological task. This would help ensure that even if some species were lost, the system might remain relatively undisturbed, as other species filled in as backups.

Part of our species protection included occasional intervention to assist individuals that needed extra attention. An example was pruning back plants which were stealing the light from understory plants (such as those morning glory vines), adjusting rainfall and temperature patterns to match the changing needs of the biomes as they matured, weeding out invasive plants, or as is the case with corals, removing macro-algae invading the outer rim of a colony's tissues. Yet most of the time, we let events take their course because it is quite natural that some species will do better than others in a new environment, and the focus of our observations is what Biosphere 2 itself is doing. Some losses are also a result of an ecosystem maturing. For example, the development of tree canopies will mean that some early developing species will eventually be shaded out by slower-growing giants. A critical issue in intervention is whether the endangered organism is vital to the ecosystem, because it is irreplaceable in the food web. In addition, some species get special attention because of their value to the humans, whether for their beauty or their utility.

Exact numbers of species extinctions will be revealed more accurately in our detailed resurvey during the transition between crews. During this resurvey of all the plants, animals, fungi, and microbiota we will also document what is new — seeds that were carried in with the soil that have now sprouted, for example, or plants that have begun to spread by establishing new seedlings. We'll also take measurements of all the plants' dimensions to determine their rates of growth and how the original stock of carbon is now distributed. It appears at present that overall losses have been mild — less than the twenty percent we had anticipated. For example, there was only one species (represented by one individual) of hard corals that was lost. We find this rather remarkable given the sensitivity of coral reefs and the notorious difficulties associated with their relocation.

In addition to the power of sophisticated sensors and analytic techniques to detect substances, there was the human organism, one of nature's most sensitive sensors. People can detect extremely small changes in temperature, humidity, and odor. A trained field ecologist is even more finely attuned to the many factors that give an ecosystem its characteristic feel. So, in a sense, we used a dual set of monitors to ensure that environmental conditions and life systems were healthy in Biosphere 2: the artificial intelligence of electronic sensors and chemical analyses plus human intelligence and sense organs.

The biospherians took regular walks through the wilderness biomes to check their overall health and the health of individual species. In many of the biomes there are organisms which serve as 'biological indicators', a sort of early warning device of emerging problems. For example, in the coral reef the colors of the giant Pacific clams and the corals indicate whether or not the organisms are healthy. In other biomes, certain plants are checked to see if their buds are swelling to help us determine the best time to activate a dormant ecosystem. We also keep our eyes on the dominant vegetation, such as the red mangroves in our marsh system, as a bellweather of system health.

✣ *Our legacy to the future*

We have been especially vigilant because of the many unknowns we faced in this initial shakedown of a man-made biosphere. With dozens if not hundreds of potential disaster scenarios, we had to be continually alert: in a small closed system, things happen with amazing speed. This speed-up of natural processes represents a good deal of the scientific value of Biosphere 2. It's like a time microscope; you can see more processes in the same amount of time.

In its first two years this new Biosphere made a good start on its long journey of growth and development. It is far different, more thriving, and with much larger biomass than the Biosphere we entered in September of 1991. The reins of operational responsibility will be turned over to succeeding crews. Albert Howard, the English soil scientist who sparked the revival of interest in organic farming, once said that the primary duty of those working the land is to hand it over to future users in at least as good a condition as they found it. This has been the ideal motivating the research department as it set our standards, to ensure that we will hand over a more mature but still remarkably diverse and beautiful Biosphere.

CHAPTER 4

Growing Your Own

It is one thing to grow some food, but quite another to live entirely on the food you grow. All of us had trained in the prototype agriculture systems we had used during the pre-closure development phase, both in the research greenhouses and in Biosphere 2. Even though we ran week-long experiments living off what the agriculture areas produced, it was not the real thing. Our experiments were brief — and the supermarket was still just a short drive away. Everything changed when the doors closed behind us on September 26, 1991. Our grocery store, pantry, and restaurant were now all shrunk to an incredibly small plot of earth. We would have to learn to make it into fields of plenty or face a long and hungry two years. A group of educated, urban Americans and Europeans were about to become subsistence farmers.

❧ *Intensive is the word*

We called our farm the Intensive Agriculture Biome (IAB). It is about half an acre of land, including the domestic animal bay and a tropical orchard. Out of this half acre came virtually everything we ate, supplemented by some bananas, papayas, and coffee beans from the rainforest and some passion fruit from the

Mark Nelson hoists a ripe papaya, one of the most successful fruit crops in Biosphere 2.

savannah. It provided food for the animals as well, so it was also the source of our milk, eggs, and meat. Half an acre is a very small piece of land to feed eight people. Even the vegetarian diets of India or China need far more land per person than we have. To add to our problems, the intensive agriculture area is covered by a series of vaulted spaceframe arches, which, even with the best of designs, cut out more than half the sunlight coming from outside. Finally, we were prohibited by our own EPA from using any toxic pesticides, herbicides, or chemical fertilizers. Under all these constraints, we had to produce a nutritionally balanced diet for eight — and sustain it for two years. We were allowed to build up a three-month reserve of food from farming Biosphere 2 before closure to tide us over the first winter. We were warned, "There are no Indians to tide you over like Mayflower settlers with a bounty of pumpkins, cranberries, and turkey, so you'd best have something to start with." This will also be true of starting a colony on Mars.

From our previous agricultural trials in the research greenhouses and in Biosphere 2 before closure, we knew it could be done. The numbers showed it wouldn't be easy — even without serious losses to insects or disease we would only just make the target. We did not know how many hours a day we would have to work on the farm to survive, either. Would the work leave us no time for our other research? There was a delicate irony here: we lived in a $150 million dollar facility with some of the most innovative technologies ever developed, yet the fundamentals of survival unite us with almost every other human since Adam and Eve. Could this high-tech, information-age system be a model for future intensive, chemical-free systems?

In order to develop a sustainable system that was going to

work, we had to take skills and knowledge not only from modern agricultural science but also from ancient intensive farming systems. An old proverb says that the best fertilizers are the farmer's footsteps — that is, it's the constant attention of the farmer that makes for a good crop. We all dreamed of a lush cornucopia, a Garden of Eden, a life-sustaining oasis. These are images far removed from the realities of twentieth-century farming. These days farmers sit in air-conditioned tractors applying chemicals to ground they will hardly tread, producing food that goes to the world's increasingly large urban population which neither knows nor seems to care how and where its food is produced. There was going to be many a biospherian footstep in the IAB trying to make the farm of our dreams and fill our plates with a satisfying quantity and variety of foods.

*⁊ Variety is the spice of life

In developing the agriculture system for the biosphere, we knew that variety would be the key. If we managed to produce the proteins, calories, and fats necessary to keep us functioning but the food was dull and unappetizing, then life would become not unbearable, perhaps, but certainly uninspiring. With some amusement one day, several of us watched a TV news story about some NASA-sponsored research at Purdue University which explained the process of food production for future astronauts. They were attempting to engineer three species of plants, rice (grain), cow peas (protein), and rape seed (oil) in order to supply astronauts with a minimum, nutritionally complete, diet. This may look good in theory, if you ignore trace elements and some enzymes, but some basic points seem to have been missed. One is that with only three crops, a failure of any one of them would mean disaster. Another is one of the more important lessons we can offer to those involved with space exploration, though we are perhaps not the first to learn it: variety and diversity in cuisine is essential to human well-being. If you consider living off the planet for a long time — not just a few days or weeks in a space capsule where the effort is focused simply on achieving a short-term goal — then you must design a life style that can support you not just physiologically but psychologically as well.

It would not have been possible to exist on a diet where all of our 2,016 meals over the two years were composed of three plants. We already have more than eighty different crops in our Biosphere 2 agriculture and will probably add more in the years to come. Just as we cannot change our genetic make up, we cannot expect humans to change their culture overnight or to survive only on proteins, carbohydrates, and fat. A healthy psychological and social life is inextricably linked to a rich and varied diet.

We hear much about wilderness biodiversity, but preserving agrodiversity is also a major issue in the world. The diversity of crops in the IAB certainly made it just as pleasing to the eye as any of the more exotic wilderness biomes. On a spring morning you would pass fields of waving wheat, some golden brown awaiting harvest, some green with seedheads just beginning to fill, others still looking like an overgrown lawn. Another beautiful sight is the 'potagerie', the vegetable patch, with a variety of plants from lettuces to chili to squash. Then there are the sweet potato patches covering the ground with vines, sorghum fields with heavily-laden heads, and white potatoes at several stages of growth. Heading along the north wall you duck under banana bunches hanging over the pathways, with slim papaya fruit ripening to a golden glow on trees in their shadows. Looking over to the right you may wave to a group of visitors peering into the windows in front of the rice that is just emerging from the mud. Alongside the paddies arch great stands of tropical lablab beans. Heading to the kitchen, you pass through the tropical orchard — a dark, green shadowy world where banana trees soar twenty feet towards the high ceiling. You will see taros (a tropical starchy tuber) with their giant fan-like leaves, and delicate citrus, fig, and guava trees which have staked out a position at middle height. Grapes, mint, and pineapples grow in the understory.

The crops chosen for the IAB had to be suitable for our tropical conditions. They also needed to mature quickly to ensure maximum use of the small space, resist disease, flourish under the reduced light levels beneath the structure, and be good to eat. In other words we were looking for the 'stars' — known, little known, and as yet undiscovered — of the agricultural world. We started with a system based on sweet potatoes, white potatoes, and bananas for starch; wheat, sorghum, and rice for grain; peanuts and goat's milk to provide the fat in the diet; and a variety of beans for the main source of protein. We also grow a large variety of seasonal vegetables as well as herbs for flavorings and fruits as sweeteners for desserts.

❧ *A hill of beans and a plot of potatoes*

Even before closure we had experienced disease problems with many of the more common bean varieties. As we went into our first winter, most of the beans in the ground were a tropical

variety called lablab beans which seemed to show some resis-
tance to disease. We had heard from Australian growers that
they should flower around the time of the fall equinox as days
began to get shorter. But the days and weeks rolled by with no
sign of any flowers, and every day we worried a little more.

Sally and Jane nervously discussed what we could replace
them with and how we could use their foliage. At least we would
get a solid supply of high-protein fodder for the animals if we
had to cut down the plants, then over ten feet tall. Finally, on
October 19, three weeks after closure during an early morning
Saturday inspection tour, Sally crackled the good news over our
radio channel: "Beautiful purple flowers on the lablabs!" Curiously enough, we also had a U.S. variety lablab planted on the
south end of the IAB where they trailed down towards the
basement, making use of the abundant sunlight that otherwise
would be wasted on the cement wall. We called it the south
foodfall. They had been there since the previous winter. Within
a day they were putting out beautiful white flowers like their
Australian cousins. A genetic program geared to day length and
light had triggered simultaneously in both of them. The lablabs
bore abundant flushes of all-important, high-protein beans for
the next six months.

Sweet potatoes were from the start one of our mainstays,
but we had much to learn about cultivating them under our
unique conditions. In the first year the crops of sweet potatoes
were magnificent to behold — but only from the surface up.
Their luxuriant leaf growth was balanced by a paucity of tubers
underground. After consulting with Dr. Phil Dukes from the
USDA vegetable labs in South Carolina, we came to the conclu-
sion that our sweet potatoes had life too easy! Plentifully sup-

plied with nutrients and water, they had no incentive to plan for the future by storing food in their tubers. We were told to make life harder on them. First we were to 'dry shock' them until they wilted. Then we were to reinforce the message by ruthlessly cutting back their long-running vines so they couldn't reroot all over the place. This forces them to send their food down to the main tubers. We were also advised to plant them twice as thickly to compensate for the reduced light. At first, all this advice was followed a little timidly. It was only after we had harvested a few plots and found that the yields were best in the driest parts of the field that we gained confidence in the new approach.

Thankfully, sweet potatoes are highly resistant to most of the insect pests and diseases we experienced in Biosphere 2. Not so with our other starch staple, the white potato. Concerned about the possibility of an outbreak of potato virus, Sally learned the technique of growing white potatoes from tissue-culture clones. Although it was labor-intensive, it looked like we would have a way to ensure virus-free potatoes in the future.

But in January 1992 we suffered an invasion by a pest we had never before seen — the tiny broad mite. First it took an inordinate interest in our yardlong beans, then moved on to an important field of white potatoes started from the virus-free tubers. Since we were unable to adequately control the minute broad mites, we had to greatly reduce the amount of white potatoes we were growing and look to other sources of starch such as taro and green bananas.

Sally divided our tiny farm into eighteen plots. The crops are rotated from plot to plot. In this way we enriched the soil with legumes (nitrogen-fixing plants) and lessened the chance of harboring insect pests in the crops or soils. Most plots hosted three or four different crops over the course of the year. Sally alternates crops that do best in intense light and high temperatures like sweet potatoes, peanuts, squashes, and sorghum, with those that do best in cooler temperatures and can manage with lower winter light levels: wheat, carrots, cabbages, beets, white potatoes, peas. Other crops can manage in either season, although their winter harvest is slowed by lower temperatures and diminished light.

Sometimes it takes a fine touch to juggle the two sets of crops — especially in the border seasons as we changed from one regime to the next. For example, if temperatures get above eighty-five degrees, wheat pollination is lowered and premature heading occurs. So a typical winter temperature control is set at sixty-five to eighty-five degrees. But should temperatures fall below sixty-five degrees when summer rice is flowering, poor seed set could result. Sweet potatoes like higher temperatures, but should temperatures get above ninety degrees, sweet corn will develop empty heads and eggplants will produce little fruit.

We were highly dependent on our electronically controlled 'technosphere' to help us with this juggling act. The temperature and humidity regimes are controlled by large air handlers in the basement. Alarms sound if temperatures above or below the safe range are recorded. This gave us time to adjust the air handlers before any damage is done to the crops. The technosphere also took much of the tedium of watering out of our hands through the use of an automatic irrigation system which waters individual plots according to the needs of the different crops. Even with all of this technical aid, frequent checks of the crops were necessary to ensure that they were growing properly.

Overwatering makes our soil sticky, and it drains poorly.

This type of soil condition would also cause excess carbon dioxide to be released into the atmosphere. Underwatering hurts the crops and makes the ground as hard as concrete. We put soaker hoses down in problem areas, often on the edges of plots, where the automatic sprayers don't evenly cover the ground. A week or so before harvests, the irrigation was usually turned off so that the field could be turned while minimizing soil compaction caused by the crew slogging through muddy soil. A programming error could (and on several occasions did) result in a group of biospherians being inadvertently soaked while weeding or harvesting a crop!

✣ *Working in the fields*

Planting and harvesting rice was new to all of us. Several of us had watched it being done by village women in India, Nepal, or China, but actually doing it ourselves was an adventure. Wading around in the thick mud, we became more and more expert at transplanting the tiny rice seedlings. Since the paddies were right by the windows, the rice harvests and plantings were favorite events for the visitors, who could watch as crew members awkwardly tried to net the tilapia fish being raised in the paddies. It's a very space-efficient system. The fish eat azolla, the water fern that grows on the surface, and in turn they fertilize the rice. The fast-growing azolla is also harvested as chicken feed. The rice takes about four months to grow, so we got about three crops a year although the winter crop is far poorer than the others. Timing here is all important. We learned to start new rice seedlings thirty to forty days in advance so they would be ready to go straight into the ground just at the time of the next harvest.

The crew always enjoyed the peanut harvest. One or two people used shovels or pitch forks to gently dig up the peanut plants. The rest of the crew sits perched on upturned buckets picking the nuts from the roots of the plants and piling them into a bin. The greens are piled up and taken down to the basement to be dried in front of the air handlers for use as winter goat fodder.

Certain agricultural crews actually required a dress code. We discovered that delicate thinning and weeding jobs in between rows of plants must be done with bare feet so that the small seedlings do not get trampled by heavy work shoes. Sorghum, not a great favorite with anybody, requires long-sleeve shirts to cover our arms and bandannas to cover our noses during harvest. This is because almost everyone has allergic reactions from contact with the leaves or spores of the heavy sorghum seedheads. Nearly all of us got into the habit of coming to IAB crews with a pair of pruning shears or a sickle tucked into our belts. Sally was rarely seen without a belt pouch stuffed full of essentials such as a lens for periodic spot-checking of insects (both pests and beneficials), a water-proof note book and a pen for jotting down observations and crop yield data, and lengths of string for tying up the occasional straying tomato vines.

However efficient the automatic equipment we had available, there was never any respite from the hard physical labor of farming. The system would not wait for us to gradually adjust to living inside Biosphere 2. Fields ready to harvest are as insistent as a bedside alarm clock. Every day we didn't work in the fields was a day lost in the filling of our breadbasket.

We were not used to so much sheer physical work — digging soil, harvesting crops, and running threshing equipment. To make matters more difficult, we had to take on the work while adjusting to a lower-calorie, lower-fat diet than we had experienced before. We soon learned to be as efficient as possible, working in large teams early in the day to accomplish heavy tasks like digging up potato fields, and allocating to later in the day the tasks that required less muscle power but more care and attention to detail. Almost everyone spent two hours after breakfast working in the agriculture system. Sally, as manager of the whole agriculture system, coordinated all the farm work. Different members of the crew took on various special responsibilities. Mark Nelson was in charge of the wastewater treatment system, the balcony planters, and the partly shaded south basement area that he turned into a sweet potato and papaya production area. Gaie took care of the orchard; Jane was in charge of the basic care of the field crops and the animals; Laser ran the compost system and raised worms in the basement; and Sally paid special attention to the vegetables.

❧ *Demands of recycling*

Not only did we have to raise all the food we needed for the two years, we had to do so while recycling all wastes, controlling pests and diseases without using toxic chemicals, and ensuring that our soils maintained their fertility. In order to maintain an agricultural system that was going to last over time — be truly sustainable — all nutrients taken out of the soil in the form of plant material had to be recycled. We couldn't use chemical fertilizers or we wouldn't be independent of such outside supplies. And there was no reason we should need them. Our agriculture, in theory, should be able to maintain its fertility if we return all the nutrients we take from it.

In many ways the animals are compost machines on legs. They happily eat many of the plant-parts that humans can not digest and turn them into manure that could be composted. The water that washes down from the animal bay and the wastewater from the human habitat fills a series of holding tanks in the basement where the digestion of the nutrients they contain is continued. Next, the water is circulated through tank lagoons that contain a variety of fast-growing aquatic plants. Here the plant roots and their microbial helpers complete the conversion of waste into valuable nutrients. Many of the nutrients were used to fuel plant growth. The new growth was harvested for use as animal fodder, and, through the animal products, was given back to us city-folk of Biosphere 2. The remaining nutrients were sent along with the water to the irrigation holding tanks to be delivered directly back to the agricultural soil they came from, and the cycle began all over again.

Composting and turning the fields causes carbon dioxide to be released into the atmosphere. During the winter months we had to be especially careful. Carbon dioxide levels would be rising and we would have to do everything we could to keep them steady. This meant that compost material had to be shredded and stored dry during the winter months, ready for composting in the spring when carbon dioxide levels were lower. We also turned to minimum-tillage strategies in our fields, sometimes just lightly rototilling the first few inches to make a seed bed rather than deeply digging the whole field. As in all aspects of life inside Biosphere 2, we had to take into consideration the effect of every action on the whole system.

❧ *Pest control*

Our pest control program was another example of how our agriculture had to be compatible with the rest of this tightly sealed environment. It would be literally suicidal to use toxic pesticides or herbicides inside Biosphere 2. With our small air volume and rapidly recycling water, we would be breathing and drinking those chemical residues within days. We had to protect our crops by a variety of safe techniques such as using beneficial or good insects to control the 'baddies' (fourteen such allies were introduced before closure), using resistant crop varieties, using safe and biodegradable sprays, washing plants with water sprays, or even picking pests off by hand. Keeping humidities low also helps control many potential pests.

We are pioneering new techniques of biologically safe pest control in Biosphere 2. Some farmers in the outside world are using beneficial insects, which they can easily obtain from insectaries. But we had to sustain our populations of beneficial insects by providing food and habitat for them when they had eaten up most of the pests. Our chief enemy in the insect battles was the broad mite. These microscopic little mites loved the hot, humid conditions in the IAB. Our attempts at control varied from the marginally successful, such as lowering the humidity and introducing predatory mites, to the completely unsuccessful bright idea of blasting them off the leaves using a portable hair dryer!

The mite saga brought back memories of the fierce pest we battled in the months up to closure — the pillbug. This critter, with its shell-like exterior, is actually one of the few crustacea (same biological family as crabs and shrimp) found in soils. Normally they function as detritivores, digesting dead organic materials from roots and plants. But occasionally they take a liking to the foliage, as we discovered in the spring and summer of 1991 when bean seedlings began to mysteriously disappear.

When we discovered the culprit, we began an intensive investigation of possible control measures. Pillbugs like moist soil, so one element of our farming had to be changed. Instead of keeping a mulch cover on the plots to keep them moist, we'd let them dry out a bit during the day to deprive the pillbugs of their cover. Although they are sometimes seen in the daytime, pillbugs are most active at night. They seem to boil out of the ground from particular underground spots where they congregate. We set out baited traps for them and even contemplated running pipes — cut in half and filled with water — as a kind of moat around plots planted with crops they particularly relished.

Then someone decided to try vacuum cleaners. After all, there were thousands of them, and since they seemed such a social lot we could start where they congregated. During several of the early week-long experiments, night crews of biospherians would vacuum beanplots by flashlight. Impressive numbers wound up in the bag, but this made little difference when we tried to get our soybeans started. The technique, ineffective as it was for controlling pill bugs, did at least provide a tasty snack for our chickens.

Our next tactic was to place plastic collars, sunk a couple of inches into the ground, around the young plants. The pillbugs must have laughed as they dug their way under our barriers. Then someone had the idea that rather than taking the pillbugs to the chickens, we could put the chickens in the field. So, in an effort that will be memorialized as one of the Truly Ludicrous Moments in Bug Control, we had chickens on tethers staked in the middle of our bean field, patrolling for pillbugs. We even had some portable chicken barriers made so that we might set up a temporary chicken run in between crops. The modest rate of ingestion of pillbugs by the chickens, and the fact that we needed to get plots replanted immediately, ruled this out as a viable option.

Finally, someone made a chance observation. Tomato plant prunings, if left on the ground and kept wet by our automatic sprinklers, were invariably thick with pillbugs. Eureka! We could put out tomato leaves as a kind of border around the vulnerable plots, thus drawing pillbugs away from the seedlings. It's a wonderful solution because not only do we keep pillbug numbers down, but we get a high-protein additive to the diet of our chickens.

The cockroaches also provided an extra snack for the chickens. Cockroaches are a species that clearly enjoyed life in Biosphere 2 as much as they enjoy it almost everywhere else. We had introduced an exotic Surinam cockroach, into the rainforest to help chew up dead leaf matter, and these stayed in their econiche. But the other, more common types managed to sneak in during construction. Their numbers began to proliferate, though at first they were mainly restricted to the wilderness biomes and basement. Soon they had migrated to the kitchen and agriculture areas. To get rid of them we tried trapping them in buckets or mason jars baited with rotten banana peels. By the second year they were getting out of hand. These large roaches decided not to stop at merely breaking down dead plant matter, which is their rightful role in the food chain. They took a strong liking to nearly every type of living plant in the IAB. Tomatoes, sweet potato leaves, and the flowers and fruits of squash plants were a particular favorite. You can imagine her surprise (not to mention horror and revulsion!) when Sally ventured down with a flashlight to find herds of cockroaches grazing on rice seedlings and nearby sweet potatoes. They even made a go at eating green sorghum seeds by climbing up the eight-foot-high stems! We scrambled to harvest ripe papayas and figs before the cockroaches found them and burrowed their way in. To try to keep them under control, half way through year two we introduced a predatory wasp which parasitizes their eggs. Also, nightly vacuuming of the front and back kitchens became part of the day watch's duties. In the morning the vacuum cleaner would be emptied in the chicken coops. The silver lining to this particular cloud was that the chickens liked the high-protein cockroaches and egg production increased.

At one point powdery mildew on our squash became a problem. Sally tried to control it by spraying the leaves with a potion of our food-grade machine oil mixed with bicarbonates used in buffering the ocean. She had read that this would be a possible control for mildew and, sure enough, for some time it did a successful job. A small example of how we had to be quite resourceful and use whatever we could find to manage a problem!

❧ *Catching the sunfall*

Other clouds loomed. Well into our first winter, we became aware of a problem over which we had very little control. Sunlight, our most precious resource, already severely reduced by the glass and spaceframe structure, was reduced even more by unusually heavy cloud cover. The normally high percentage of sunny days is what made us select Arizona as the location for Biosphere 2 in the first place, but in this year abnormal weather patterns worldwide (this was the year of El Niño) caused excessive rain and flooding in Arizona and an unusually high number of cloudy days.

Each successive cloudy day brought rising carbon dioxide levels and more problems with the development of the food crops and their susceptibility to disease. Every bit of sunshine was precious, so a process was begun which stayed with us throughout the two-year experiment. We called it 'catching the sunfall'. This involved taking every possible space we could find in the agriculture area that received sunlight, no matter how small an amount, and catching the sun there with plants.

Mark Nelson was one of our chief 'sunfall catchers'. He started by cramming as many pots and plastic jugs full of soil as possible into the south basement area, using them for an extra crop of sweet potatoes. After that, he and Laser teamed up to create additional planting boxes using artificial lights that were no longer needed in the algae-scrubber tanks. They scrounged up wooden planks or cinder blocks to make the planting bed, hauled the soil in, and hooked up the lighting systems. Under these lights, red beets, white and sweet potatoes, squash, and even radishes flourished. Planting boxes sprouted in the IAB basement along the sunny south wall. There were even two in the plaza outside the IAB doors where we took our breaks. One day, when several of us were sitting and chatting over breakfast, Gaie told us that the previous night at three o'clock in the morning she ran into Mark who was gingerly moving a large lemon tree down the hallway on a dolley. Mark, subject at times to spells of insomnia, was working into the night to cram the balcony full of plants. Soon the balcony, which receives some of the best light in the IAB, was home to thirty new papaya trees and 400 running feet of planters with lablab beans and sweet potatoes. It became a running joke that Mark would eventually plant up our personal rooms or put floating tubs out in the ocean to soak up more light. But it was also deadly serious; we needed more food producing space — especially if the low light levels continued.

That lemon tree began the transformation of the balcony overlooking the IAB. We had grown sugarcane and leuceana trees up there after closure, hoping their fast growth would give us a boost in carbon dioxide management. But the sugar cane produced a lot of green stalk and not much sugar; so Roy had taken over the balcony to grow beans. Impressed by how fast papaya trees come into bearing (nine months after planting you can harvest your first fruit) and with the quantity of fruit they produce (up to hundreds of pounds on a fully productive tree), Mark began to interplant papayas and bananas next to the beans on the balcony. By the time he was done, there were twenty-five more fruit trees on the balcony. There in the full sun we ripened papayas to use as fruit, while in the south basement area, with its poorer light, we harvested the papayas green as a vegetable.

From there, Mark continued his expansion by constructing a line of planters all around the front part of the balcony terrace. The light was excellent, and the beans and sweet potato vines could hang down over the railing to catch some of the light that presently fell unused. We organized bucket brigades to winch up soil from the IAB to fill the low containers and pots. We transplanted fig, lemon, and orange trees from the orchard, where they were doing very poorly in the low light, to the space around our café tables where we usually had our Friday dinners.

The process became a meditation on space and light. Suddenly it was possible to see new patterns and opportunities. One evening Mark discovered a new pet project. He was idly looking down on the IAB with Laser when they had another 'food flash'. "The stairwells! We can plant up the stairwells!" The IAB had four large stairways. The ones on the north were shaded by the wall of bananas and papayas, but the one on the south was hardly ever used and the one we did use could be boarded over. The two spots had excellent sunlight and would add an additional 250 square feet of planting space! The project was rapidly put into motion with Laser engineering the main structure and everyone else helping to move soil and fill the planters.

✤ Dialogue of biospheres

By the second fall we really felt that we were coming to grips with the challenges of food production and were starting to understand exactly what it took to create a completely self-sustaining agricultural system. We also felt it was time to invite outside consultants to join us in this incredible venture — especially where we needed additional expert advice.

Sally and Mark organized an agricultural workshop so we could share our data with others in the field and try to interest a top group of agriculturists to collaborate on long-term research just as we had done with the marine and terrestrial systems.

The key person we discovered was Dr. Dick Harwood of Michigan State University. He was recommended by Eldor Paul, the soil scientist from Michigan State who'd attended our April 1992 Closed Ecological Systems workshop. "Dick is the man you're looking for. He's a field agriculturist. Knows the science and is practical as well." Indeed he is. He has served as coordinator of Asian programs for the international aid group, Winrock International; directed the research program in organic farming for the Rodale Institute; and served as ecological consultant to the International Rice Research Institute (IRRI) in the Philippines when they decided to use an ecosystem approach rather than the chemical path of the Green Revolution. Currently Dick holds the Chair of Sustainable Agriculture at that previous bastion of chemical agriculture, Michigan State University. He is also involved in the landmark National Academy of Sciences' reports on Sustainable and Alternative Agricultural Systems which are challenging our current policy of subsidizing and encouraging expensive chemical agriculture with its attendant serious environmental pollution and high costs.

Dick in turn led us to Dr. Litsinger, an entomologist with extensive tropical fieldwork behind him. Jim worked in the Pacific Islands during his Peace Corps days, then went to IRRI to consult on biological control strategies for rice and other tropical crops. From Winrock came Dr. Will Getz, an expert in small animals who consults in tropical African development projects.

Finally in our agricultural workshop we had found our natural allies: experts who could see the relevance of Biosphere 2 to the accelerating shift all over the world away from chemically-based agricultural systems, and people with the practical skills to enable us to integrate what was being learned elsewhere. Dick was especially intrigued with the possibilities.

Nestled in the ridges of the Santa Catalina Mountains, Biosphere 2 is the largest, most tightly sealed ecological system ever created.

Mission One crew (from left): Taber MacCallum, Mark Nelson, Sally Silverstone, Jane Poynter, Linda Leigh, Abigail 'Gaie' Alling, Roy Walford, Mark 'Laser' Van Thillo.

Co-captain Sally Silverstone presides at a morning breakfast meeting.

Jane Poynter (left) feeds the goats a morning meal of peanut greens, as Sally (right) milks.

*Linda Leigh (left) revels in an
abundant peanut harvest,
while Mark Nelson (right) gathers
sheaves of rice in the Intensive
Agriculture Biome (IAB).*

Mornings began with a two-hour stint in the fields.

Sally uses a magnifying lens to identify pests on the leaves.

Bananas are the most reliable starch and most appreciated sweetner in the Biosphere 2 diet.

Laser forks inedible leaves and stalks into the compost bin.

Taber and Jane take their mid-morning break on the patio overlooking the IAB.

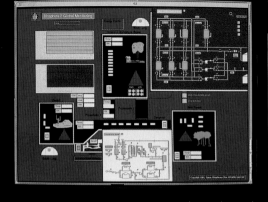

(above) Taber MacCallum runs tests in the analytical lab. (right) Nearly 2,000 sensors throughout the Biosphere register data on computer screens inside the structure as well as outside in Mission Control.

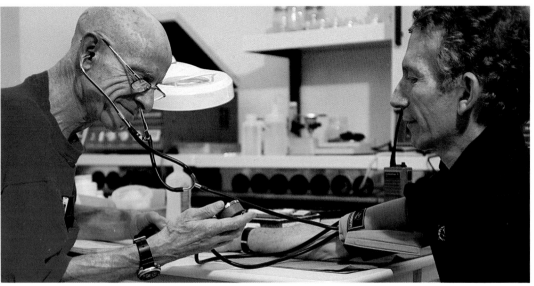

Dr. Roy Walford checks Mark Nelson's blood pressure during his bi-monthly medical work-up.

*(top) The command room houses the crews' individual
workstations as well as serving as communications hub.
(above) The windows in the dining room look out
over the verdant agriculture area.
(right) Two lobsters that outgrew their niche in the
ocean will fit very nicely into a cooking pot.*

ABIGAIL ALLING

ABIGAIL ALLING

(above) Jane, Mark, and Sally collaborate on a party meal.
(left) Linda is the queen of the coffee pot.
(below left) Taber and Gaie fill their plates with a hearty lunch.
(below right) Laser creates his special banana-papaya crêpes.

PETER MENZEL

ABIGAIL ALLING

An expert in the history of agriculture, he sees that we are approaching a watershed as we leave the fossil-fuel subsidized agriculture behind and forge new types of sustainable systems. How to do so while preserving a wide bio-diversity of crops, lowering labor requirements, and producing enough to remain economically viable is the challenge. Biosphere 2 is set up to work on that program while offering the research opportunity to study in detail how such a system operates and impacts the rest of a biosphere. As Dick told us: "The watchword in agriculture is becoming 'containment of nutrients.' What you have in Biosphere 2 goes beyond that — it gives us the chance to study the whole recycling even of those nutrients that are put into the air or drain through the soil."

The steps our new agriculture team was instituting to improve our production lifted our spirits during the dark, dark days of winter. Jim Litsinger had immediately gotten onto the case of our continuing nemesis, the broad mites. He located four new predatory mites which were soon introduced on a scientific import. One radio station distinguished itself by trumpeting "Biosphere 2 Introduces 30,000 Mice." It was eventually forced to retract that arresting story — and presumably cancel the order for the Pied Piper. It was *mites*: 30,000 predator mites fit in two small test tubes! With the counsel of Will Getz, the research staff in the BRDC also started looking into animals such as Muscovy ducks, which might do well inside the Biosphere and be introduced for Mission Two.

Dick Harwood extended our soils research by working with colleagues at Michigan State University. An expert in Chinese farming, he thought the best parallel to Biosphere 2 was the intensive agriculture of Szechwan Province. Monsoonal conditions there parallel our reduced light. He showed us slides of a farm he'd visited there where nearly one hundred crops were raised on less than three acres, with a range of crops adapted to the conditions from the steep hill-side slopes to the tilapia fish and rice cultivated in the ponds and paddies at the valley bottom. Dick began a worldwide search for other areas with similar light conditions. He also searched for shade-tolerant crops that would boost our productivity. The matches included monsoonal areas of the Philippines, Kenya, and the Ecuadoran rain forest. During transition we would further sculpt our IAB so that new crops from these areas would help us match the varying light conditions.

We could appreciate the importance of shade-tolerant plants. Our second winter was even darker than the first. Though it wasn't an El Niño year, Southern California and Arizona experienced record-setting rains and floods in January and February of 1993. Our light was twenty percent less than the poor levels of our first year! It was quite a shakedown test for our agriculture system!

❧ Getting the hang of it

Sally continued to perform miracles by juggling our food supplies. We had to dip into seedstocks to make it through, but even so, we felt successful: we didn't need to import any food during that long, dark winter of our second year.

Finally, starting in March 1993, we were really beginning to see an improvement in our crop production. The sun finally came out and gave us the glorious, long, sunny Arizona days that had led us to choose this location for the Biosphere. Even some of the bean varieties that had been impossible to grow for a year and a half started to do well. The sixty new papaya trees planted during our sunfall harvesting campaign started to thrive, and hundreds of pounds of extra beets and sweet potatoes came in from our makeshift planters.

As we had suspected, it took the reality of being dependent on our agriculture to make successful farmers of us. What a difference when we understood at the stomach-level: if we don't grow it, we don't eat it. We had learned a lot from the experience, truly learning by doing, and we eagerly looked forward to the first transition phase where we could make alterations and improvements. Of course, as in all farming systems, new problems will continue to come up. Learning how to deal with them will be part of the fascinating process of creating a sustainable and non-polluting agriculture system.

CHAPTER 5

Hunger and Resourcefulness

We took great care to see that each member of the crew got an equal portion of food. We unanimously agreed that everyone should receive the same quantity regardless of age, sex, or work regime. Organizing it any other way would have been almost impossible and probably would have caused a great deal of resentment. In the early months after closure, the cooked dishes were put out in large bowls on the counter top for people to come and help themselves in buffet fashion. However, as time went on and the crew got hungrier there was a tendency for the early birds to overfill their plate so that anyone who was a little late for a meal would be left with very short pickings. By the end of the first six months, the cooks were dividing the food into eight equal portions before serving it or using serving dishes where the division into eight portions was obvious.

Before the eight of us entered the Biosphere we got together and put on a satirical skit about how we envisioned life in Biosphere 2. One of the funniest scenes depicted the crew going down to the storeroom one at a time to steal food when hunger set in.

After closure, life began to imitate art, and this time it wasn't so funny: one or more biospherians were stealing ripe bananas from the basement storeroom. At first, Sally decided to

Sally Silverstone carefully weighs out the food for the cook of the day.

ignore the thefts, figuring that the bunches of ripe and semi-ripe bananas were almost irresistible to a hungry stomach. However, over the year it slowly got out of hand. On one occasion sixteen ripe bananas that she had earmarked for the day's allotment of food disappeared before she could harvest them and take them to the kitchen. It was then that she decided to hang the ripening bananas in a locked storeroom. The processing room freezer also became an overwhelming temptation to at least one member of the crew. When Sally discovered that her precious frozen food supply (bananas and other foods) had been hacked into by a person or persons unknown, she decided that this too had to be locked.

❧ Making plants taste good

You know that they make bread with wheat. And you probably have some idea what a field of wheat looks like. But do you have any idea how those plants become bread? Most people have at best a vague notion of how such foods are actually processed to make them edible, but in the Biosphere, we needed more than a vague notion. Planting, growing, and harvesting was only half the story: we had to make food out of what we harvested.

Finding suitable equipment wasn't easy. We operated on too small a scale to use machinery designed for large farms or food processing plants. On the other hand, we had to minimize our labor with some kind of machinery, and we needed equipment for a wide variety of crops. Some, like sugar cane, are not common in the United States, so it took quite a search before we found a good simple sugar-cane press for squeezing the sweet liquid from the stalks.

Wheat, sorghum, and rice — our principal grains — can't be eaten without considerable processing. When we harvested our wheat, we then cut the sheaves in handfuls with a pruning shear or sickle, stacked them in bundles, and laid them into large carrying buckets with the seed heads pointed out. Then we carried them to the basement of the IAB, where we kept the equipment to make the stuff edible.

After drying the wheat for several days in the large drying ovens, we started threshing. The threshing machine was a little small for the job, and could be dangerous if operated by an inattentive biospherian. Soon after closure we got a vivid demonstration of the dangers. Jane had assigned Roy to thresh rice down in the basement of the IAB. He got the machine started, and began feeding rice into it, but yelled for help when he couldn't get it to work properly. Jane, working nearby, charged into the room.

"Look! If the stalks build up inside the machine, it won't be able to thresh the rice. That's the problem," she said, suspecting that Roy was looking for an excuse to pass the job on to someone else. She put her hand into the slow moving barrel to pull out the stalks. "Oh God! Roy, turn off the machine — my hand is stuck! Turn off the machine!"

After what seemed like an eternity to Jane, Roy turned off the power and Jane lifted her hand. The top part of her middle finger was nearly severed. Roy took her to the medical room, where he and Taber tried to sew it back on. Roy finally decided that Jane should have the benefit of specialized hand surgery and insisted that she go outside the Biosphere for medical help.

So we tried several alternate ways of threshing the grain. One was to lay all the wheat on a tarpaulin and have the entire team stomp on it with bare feet. This was fun initially, but we soon realized that it took more time and was less efficient in removing the grain from the hull. Concerned about loss of time, Sally and Linda — the most efficient threshers — took over the threshing work. They soon developed their own methods of making the machinery work. Threshing away in the dimly lit basement was rather like digging for gold in a deep mine — the pleasure came in watching the loot pile up in the buckets and then weighing the final grain harvest.

After threshing, the grain is winnowed to blow out any remaining husks and then ground into a coarse, whole-wheat flour ready to make breads, pie crusts, cakes, pancakes, crêpes, or porridge. We stored the straw from the wheat for use as domestic animal bedding. This ensured that any grain that escaped threshing would be easily found by the chickens or the goats.

Experimentation led to discovering the best way to process dry beans. Eventually we hit on the idea of stuffing the beans into a burlap sack and pounding it with a rubber mallet. This would break open the dry shells and then the beans could be winnowed out, ready to store or cook. This is a particularly good job for someone who needs to take out his aggression on an inanimate object: banging on the sack, under the right circumstances, could be a therapeutic activity!

We were less successful with the dehulling of rice. Sally had clear memories of the village women of Bihar, India spending a considerable amount of their waking day pounding the rice with a large stick in huge wooden pestles to remove the hull and then winnowing the grain and hull mixture by throwing it up into the wind over large woven mats. Clearly we would not have time to do this in Biosphere 2. Eventually, we obtained a small Japanese dehulling machine, but it would continually jam if even the smallest piece of straw got caught in it. After many frustrating hours spent taking the machine apart to unjam it, Sally developed a method of operating the machine with one hand while holding it together with the other. This way when it became jammed it was easier to open it up and clear it out.

Once dug up out of the ground, sweet potatoes had to be cured by leaving them in a cool, dry place for several days before they could be stored. When the fields were dug up everyone would try to guess how much we had harvested. Between the eight of us, we ate approximately ten pounds of starch a day, so the big question was always how long a particular crop would last. A good harvest meant sighs of relief all round, and maybe a few extra sweet potato pies. A poor harvest would send Sally on an inspection tour of the banana trees and taro plots to calculate how much starch we could get by eating them until the next sweet potato crop was ready.

Fruits and vegetables were harvested daily and used as part of the day's cooking allowance. For example, we cut banana bunches from the tree when they filled out and a few bananas showed signs of yellowing. Later, the rest of the banana tree was cut down and composted, its leaves carted to the animal bay for fodder. The green banana bunches are hung in the basement and left to ripen. Since bananas were the main source of sweetness in our diet and they far out-produced our ripe papayas, figs, and guavas, they were carefully rationed out so we could have some every day. When a whole bunch of bananas ripened at once, they would be broken into small pieces and frozen in three-pound bags for daily cooking allotments.

✺ *Cooking and eating*

Once every eight days each member of the crew cooked for everyone. The cook's 'tour of duty' would start with the evening meal and continue over breakfast and lunch the next day. This way soups could be left to cook slowly overnight in ceramic crockpots, and cooks could organize themselves so that the bulk of the cooking was done the previous afternoon leaving them free to concentrate on other tasks the next morning.

Sally would carefully weigh out the day's allotments of fresh foods for each cook. Dry foods, such as beans and flour, were weighed out on a weekly basis and left in the kitchen for the cooks to take their allotments. A board in our back kitchen displayed the changes in daily allotments of starch, flour, and beans, depending on what was available from recent harvests. This helped us keep very accurate records of the amount of food that was divided up. Sally would then enter this information into a computer database that was programmed to determine the amount of nutrients we were taking in. This would ensure that we were keeping above the recommended intake levels of calories, proteins, and fats.

Most Americans have to watch carefully to make sure they don't eat too much fat, but our diet was only ten to fifteen percent fat — the low end of the range. Peanuts supplied most of this fat, though a small amount came from milk, eggs, and meat.

Feasting very rapidly became an important part of biospherian life. The joy of sitting back and stuffing the belly was a welcome contrast to the daily experience of being hungry after every meal. Each feast had to be thought about well in advance so Sally could save special goodies like milk, eggs, figs, and papayas. By the time a feast came around we could really prepare a spread.

Coffee was such a precious commodity that decisions about its consumption were a matter of some gravity. Usually we would wait until we had enough for eight cups, then discuss when we would drink it. Linda, whose love of good coffee led her to be particularly interested in its production, became the doyenne of coffee affairs. She would announce how many 'coffee pots' worth of coffee we had in storage and would give projections telling us when we would get our next harvest. There would then follow a serious discussion about exactly when the next pot of coffee would be brewed and which special occasion it would be used for. Linda also undertook the peeling and roasting of the beans. The night before a 'coffee breakfast' the smell of freshly roasting coffee would permeate the habitat.

Each member of the crew had a different cooking style. Some cooks spread their ingredients out all over the kitchen counters and then spent a long time pondering how to divide them up into various dishes. Others would first wash and chop everything and then play it by ear, creating dishes as they went along. Some cooks preferred to make a heavy dinner, which meant that lunch would be a little lighter. Others would arrange things the other way round. The trick was to have enough for all meals.

Breakfast is very important, especially since we do most of our heavy agricultural work in the morning. It was unanimously agreed that breakfast should always start with a large bowl of porridge sweetened with banana and other fruit when available. Some biospherians occasionally blended in cooked sweet potatoes, green bananas, green papayas, yams, or plantains. Side dishes with the porridge varied. Either the porridge would be accompanied by a sweet bread (for example, a sweet potato loaf with banana and figs on top), baked sweet potatoes, beans with chilis, or a small glass of banana milkshake, made, of course, with goat's milk.

A typical lunch menu might include a sweet potato and vegetable soup, sautéed vegetables, salad, sliced beets, and a wheat roll. Most evening dinners were vegetarian. A typical dinner could include a bean and potato burrito with hot sauce, steamed vegetables, salad, and banana ice cream. For Sunday dinners we enjoyed some meat from the chickens, fish, goats, or pigs. Aside from the three meals a day, there is a daily snack of roasted peanuts and mint tea.

The cook of the day served dinners between 6:30 and 7:00 AM. Breakfasts are served at 8:00 AM on weekdays and a leisurely 9:00 on weekends, while lunch was served daily at 12:30. Breakfast was always held in the dining room which has a view of the upper savannah. As we ate, Sally would call the morning meeting to order and each member of the crew reported what they would be doing that day. At the end of the meeting any remaining porridge was shared out by the cook. Lunch was also held in the dining room, but dinner was a moveable feast. The balcony overlooking the IAB became a favorite place to eat in the evening. This area has a café like atmosphere and became known as the Café Visionaire. We decided that our Friday night dinner would be held there. Everyone enjoyed sitting back at the end of a long week to witness the spectacle of the sunset over the Catalina mountains through the spaceframe as we ate. The Café Visionaire was also the location of some of our holiday feasts and special breakfasts. But lunch there is out of the question: during the hot part of the day temperatures frequently rise above 110 degrees.

Even though we increased the amount of calories per person after the first six months, and the weight-loss stopped, it was clear that a calorie-restricted, low-fat diet without high-sugar snacks took a lot of getting used to psychologically — especially for a crew who had been accustomed to the variety of sweets and munchies available in an American supermarket. The only snacks we had were after our two hours of agricultural work. In the first year, breaktime snacks were either a ripe banana or roasted peanuts. In the second year all the ripe bananas were needed for cooking, so our snack consisted of a handful of roasted peanuts and, occasionally, some sugar cane stalks to chew on. Other than that, food was not available between meals.

Biospherian adaptations to the food situation varied. Mark came to be known as the human goat, because he would happily snack on anything green that he could find to chew in the agriculture area. His favorite snack was the sweet leaves of our fennel bushes which he described as 'biospherian chewing gum'. It took Sally, who was in charge of the herbs, many months before she realized that the mysterious grazer of her fennel plants was a human. Both Mark and Sally, the two iron stomachs of the crew, took to chewing on empty peanut shells or even the occasional ripe banana skin to supply at least the illusion that food was being taken in.

Sally squirreled, saving little pieces of this and that from her meal plates and putting them away in the freezer for a time when she was really desperate for a snack. She actually rarely delved into her store — it was more as if she were comforted by the knowledge that it was there. Other members of the crew would save parts of their meal to eat later, especially if they knew they were going to have a late night or an especially demanding task to complete. Mark tried saving his dessert for late night snacks while Taber, Laser, and Gaie would eat everything at once, rarely saving anything for later.

Probably because we're mostly on a vegetarian diet and our ability to digest meat has decreased, some of the crew had problems with gas attacks after a heavy meat feast. Linda and Jane seemed to have greater problems with eating meat. By the second year, they had decided to decline even the small amounts of meat we had available, so cooks prepared alternative vegetarian dishes for them on our rare meat nights.

All the biospherians had to learn to be adaptable and creative cooks, since the types of food available varied greatly during the year. Sometimes the main starch would be white or sweet potato, other times it would be plantain, green banana, or taro and eggplants. In the winter we had especially heavy harvests of root vegetables such as carrots and beets.

Beets consistently did well in the IAB. Consequently, we often had more of them available to eat than we would normally have included in the diet. Our Russian visitors always laughed when we told them how much borscht we ate. Beets froze fairly well, but tasted much better fresh. In response, the crew devised all sorts of way of cooking and serving beets. We had a dozen variations of borscht (beet soup), beet casseroles, and beet salads. Someone even tried to put beets in the porridge, but this was vetoed by the rest of the crew. Gaie invented a surprising and delicious cold beet juice with lemon and chili. Laser made a beet, sweet potato, and banana milkshake for dessert one night and tried to pass if off as raspberry sherbet. We even joked that it would be a good idea to make beet wine, but this never happened.

As the experiment progressed we became quite good cooks. This is one area where intense peer pressure worked near miracles: even the worst of our pre-closure cooks became reasonably competent. Each cook developed specialties. Laser was famous for his pastries and Belgian crêpes which became a favorite for feast and birthday breakfasts. Gaie made superb ice cream and buns. Taber would cook up individual pies and Mexican pastries in all sorts of strange shapes and would create colorful and artistically arranged individual plates. Mark brewed up delicious vegetable soups, containing everything edible he could lay his hands on. Jane invented a wonderful posole recipe that was ideal for using up miscellaneous parts of the pig. Linda was known for her soups, including her cheeseless 'cheddar cheese' soup. Roy made a superb chutney and Sally became chief birthday cake baker, resident brewer of alcoholic beverages, and maker of specialty items such as goat cheese, cream cheese, and pork sausage. Eventually, Sally's reputation grew on the outside, until she was asked to do her cookbook, *Eating In* — which has been a great success.

Because of the intense interest everyone expressed in each meal, cooks became extremely careful. After a couple of incidents in the beginning months it was unheard of for a dish to be burnt. This was especially important not only because of food loss, but because the smell and smoke linger for a long time in a closed system. There were, of course, the failed experiments. For example, one cook tried to perfect a cold green soup for lunches by blending up Swiss chard and beet greens. After two tries, we all agreed that this was one experiment we didn't want to be subjected to. One time Gaie and Sally got it in their heads to cook up marsh canna roots. They carefully washed and scrubbed the slimy mess that they had extracted from the bottom of the marsh and then boiled and seasoned it. This turned out to be a disaster since they were almost entirely composed of fiber. Even hunger has its limits!

We continuously experimented with different types of food. One of our biggest challenges was making taro edible. Several members of the crew were sensitive to the oxalic acid contained in the root — it stung their tongues and upset their stomachs. It was certainly not a taste we were used to. We discovered that by peeling the outer layers of the root, boiling it several times and for several hours each time, discarding the water and then mixing it with other foods, we could make it palatable. Gaie hit on the best idea for preparing it: after doing all of the above, she then baked it for three hours until it became a crispy, golden-brown potato-like patty.

Hiding bland foods in other dishes became second nature. For example, we had plenty of green papayas harvested from trees growing in low-light conditions. They would take too long to ripen, but they proved to be an excellent starch supplement. They are quite tasteless but are useful in thickening up soups and sauces, and in small quantities are almost undetectable in a hot, morning porridge.

In the winter when vegetables were in short supply, we tried adding sweet potato greens to the daily diet. These are a staple food in some cultures, but some members of the crew had a hard time with the taste. Some cooks refused or 'forgot' to serve them, while others happily gulped them down along with everything else, appreciating them as stomach fillers. But when other harvests kicked in no one complained when they were dropped from the menu. This left more for the goats, who so far as we know never complained about them.

Plantain presented a problem at first as no one was really experienced in cooking it, but we soon learned to bake it and mix it into stews and stir fries. At first we didn't know what to do with long, thin, sweet potato tubers (some as small as a pencil diameter). We finally discovered that carefully baked they come out with a tasty, crunchy texture. Potato chips!

But gradually our tastes began to change; our cravings for chocolate and junk food decreased. We found new ways of bringing out the best qualities of vegetables and fruits. On the whole, by the second year the crew agreed that we had really excellent food, and at times we all contemplated the fact that we may never eat so well again. After all, we knew exactly where our food came from and there were no preservatives or extra sugars and fats thrown in. Ours was truly a natural and organic diet. The only problem was that there never seemed to be enough of it. We never did learn who was stealing the food.

CHAPTER 6

The Doctor Is In

In Biosphere 2, the doctor is always in. Every eight weeks, each biospherian was scheduled to meet with Dr. Roy for a full medical check up which included a routine check of weight, pulse rate, blood pressure, body temperature, eyes, ears, throat, and reflexes. In addition, Roy would take samples of urine and blood for chemical analysis. This way he could check our cholesterol and our red and white blood cell counts. Every few months, Roy took a Polaroid photograph of each biospherian without clothes on as a documentary record of the physiological changes that occurred with our unique calorie-restricted diet and associated weight loss. Every six months, Roy ran an additional series of exams which included an x-ray of the chest and an electrocardiogram to check heart condition.

The Biosphere 2 medical facility is designed to minimize byproducts that cause environmental pollution, which would contaminate our tightly sealed world, so no chemicals that could harm the air and water systems are used for tests. We also tried to minimize the time required for medical diagnosis and treatment by using state-of-the-art technologies. For example, Kodak donated one of their Vitek systems, a device for analyzing bacteria. The machine was originally developed for use on NASA spacecraft, to identify the type of bacteria responsible for

Maintaining health and vitality for the Biosphere 2 crew was top priority.

common infections. Indeed, some of what we are learning while operating our medical facility may be of value for remote human habitations such as in Antarctica or on a space station or lunar colony.

Roy is a graduate of the California Institute of Technology and the University of Chicago Medical School. He has been a professor of pathology at the University of California as a researcher in genetics and nutrition for over thirty years. When he decided to join the first two-year crew, he wondered if he really wanted to become a general practitioner, a 'country doctor' as he put it. But in the end he took refresher courses in general medicine to get licensed as a medical practitioner in the state of Arizona so that he could provide a full range of services for the other seven crew members for the two years. These courses covered dentistry, general health, prescription of pharmaceuticals, and gynecology. Special care was taken to prepare for possible accidents, allergies, and infections. SBV sent Taber, his assistant, through a special, intensive six-month training program in general medicine — including dentistry — at the University of Arizona Medical School. He and Roy had also completed a program in microbial medicine at UCLA so that if Roy required assistance, or was the patient himself, Taber could carry out the necessary procedures.

Roy's initial concerns about becoming our country doctor and being weighed down by endless complaints from sick biospherians proved groundless. As it turned out, we rarely got sick. Because we were removed from contact with germs in the outside air, what we went in with on September 26 is what stayed with us for the two years. Even our fears that we might pass around a few 'bugs' early on in the experiment were not realized, perhaps because we had been working and eating together so closely for months before full closure.

After we started to import and export research samples in July of 1992, we noticed that the biospherians who had entered the airlock sometimes became slightly ill two to four hours later. These illnesses seemed to match those on the outside. It seemed like we were highly susceptible to the bugs on the outside, and we began to wonder about what would happen when we emerged in September of 1993. It is possible that we will have become less immune to the common colds and flus because we have been living in such a protected environment.

Since the Biosphere is full of plants producing an array of biochemicals and pollen, we thought allergies might become a problem. If some of these irritants had not been taken out of the air by the filters in our air handling units the allergy problem could have been much worse. Jane was treated for her allergic sensitivities before closure, and Linda received a series of shots for her allergies from Roy during the two-year period. Even so, she still suffered a bit when the savannah grasses were in flower.

Some of the other medical problems tackled during the two years included sporadic allergy attacks, mild episodes of insomnia, stomach-aches (especially after early feasts when we hadn't learned to restrain ourselves in front of a heavily-laden table of food), and accidents. Roy handled all our health needs with one exception: Jane had to leave the Biosphere for five hours, after the threshing machine accident that severed the tip of one finger. Her finger had to be surgically closed at the University Medical Center in Tucson. Other minor accidents included the odd sickle cut and acacia or cactus thorns embedded in feet. Minor remedies such as band-aids and aspirin were always kept in a little metal tray just inside Roy's medical office.

Roy's primary research interest covers the effects of nutrition on human physiology. Many times in the past year he has said that destiny played a hand in picking him as the doctor for the first team of biospherians because one of the most interesting medical aspects of our two years inside Biosphere 2 turned out to be our diet. For over thirty years Roy has been studying the effects of a high-nutrient, low-calorie diet on mice. He found a variety of physiological effects including lowered cholesterol levels, lowered blood pressure, an enhanced immune system, a dramatic slowdown of aging, and an increase in their life span. In sum, these mice were healthier than the mice that were eating a high-calorie diet.

A low-calorie, nutrient-dense diet is exactly what we wound up eating inside Biosphere 2. So for the first time, research on the health implications of a human group eating this type of diet could be studied — and in a situation where cheating was impossible. Roy could be absolutely sure that what we produced is what we actually consumed — there could be no quick runs to the convenience store.

In the first few months we averaged about 1,700 calories per person per day. This was quickly raised up to 2,000 and then maintained at around 2,200-2,300 calories. It was nutrient-dense in that we ate grains, beans, and lots of vegetables — sugars and refined foods were non-existent, and fats minimal. The diet contained plenty of protein and enough of the other nutrients for good health. But there was a definite limit on the amount of energy-producing calories we had — and a whole lot less than we were accustomed to. Each biospherian responded differently to the diet. Initially, over the first six months or so, we lost between eighteen and fifty pounds each. Mark remembers around Christmas of the first year looking at his weight on the scale and projecting that if that rate of weight loss continued, he'd weigh minus ninety pounds at the two-year point! But during the course of the experiment, after we had adapted to our new diet, most of us either maintained our new weight, or managed to gain back some of what we had lost. Others never fully adjusted, and were either losing weight or staying at a border-line level. Roy continued to assure us not to worry when we commented on our baggy pants and loose shirts because our overall health was actually improved by the combination of our diet and the superb freshness and quality of the organically grown food.

Roy was intrigued by the wide variation in our physiological responses. We ranged in age from twenty-nine to sixty-seven, differed in ethnic backgrounds, and were a mixture of metabolic types from lean and thin to stocky and chunky. But each ate almost the exact same foods and number of calories during the two-year period.

By the time two months had passed all eight of us had had our first medical check, and we were all astonished by the results. We started with normal, healthy levels of cholesterol averaging just under 200. Now we were averaging less than 125 — the levels normally only seen in young children with their years of fast food and high fat consumption ahead of them. These low cholesterol levels lasted for the duration of the experiment.

Roy the scientist was happy at the research opportunity.

Roy the person experienced hunger like the rest of the crew. Hunger became our constant companion, always there to struggle against. Part of it was the painful adaptation from our high-calorie, high-fat pre-closure diets. Part of it was no doubt adjusting to the reality mankind has lived with for most of our evolutionary history. In between meals we were almost always hungry. We came promptly and eagerly to every meal, motivated both by hunger and the desire to get a fair share of the meal served. We often left meals, except for feast celebrations, still somewhat unsatisfied. This is part of our biological heritage — and unfortunately a reality for many people in the world today. Unlike many of the people in the developing countries who are both undernourished and malnourished, our diet is dense in nutritive value. Still, being so restricted in calorie intake for so long a time was a new experience for us.

This was the first study of humans on this type of diet, and Roy published his findings in the *Proceedings of the National Academy of Sciences* in December 1992. Since we have shown the exact same short-term physiological changes as the laboratory animals, it is intriguing to speculate on whether staying with such a diet would lead to age retardation and life extension like Roy had found with the mice. We like to think that we have grown younger instead of older during our two years in Biosphere 2, but there is as yet no way to prove that.

Many of the biospherians noticed that out of necessity they were changing old behavior patterns. Frenetic activity was out of the question on our restricted diet. No one ever heard anyone run up the stairs, and for the first time in years, many of us found ourselves taking a nap during the midday siesta period. The finite supply of food definitely curtailed useless action. There was no need to do anything in a hurry, nor could we afford to burn calories on non-essential activities. On the other hand, we were not sedentary. Most biospherians work two to three hours a day in the agriculture area, often doing manual jobs like planting or harvesting, as well as additional hours in their own particular management areas. Most of us found it hard to believe we could sustain active lives on such a reduced diet.

There are some other fascinating side-effects of our diet. For a while friends and family commented on how pale we were becoming. This was not surprising, since we were deprived of ultraviolet radiation which is responsible for tanning skin. But then we noticed that our skin — especially the palms of our hands — began turning more and more orange! This was a result of taking in high levels of beta-carotene, a nutrient found in red beets, sweet potatoes and carrots. Since these three vegetables were always in good supply in our pantry, we would walk out of Biosphere 2 with a markedly orange complexion.

Gaie and Roy were the only ones who had dental problems. About six months into the project, Gaie noticed a filling had fallen out of one of her teeth, and about a year later Roy had a similar problem. Luckily, Roy had stocked a dental paste for emergency fillings. Roy's emergency dental paste had two ingredients which harden when mixed together. The hardening takes only two minutes, so it has to be applied quickly. Roy successfully patched Gaie's missing filling and Taber did Roy's tooth. This was a temporary fix — Roy cautioned that the long-term solution would have to be found when we exited.

❧ Declining oxygen —
the slowest mountain climb in history

The unexpected decline in atmospheric oxygen prompted another kind of medical research. It declined from 20.9% — the same as in Earth's atmosphere at sea level — to approximately fourteen percent. When the oxygen levels in the atmosphere reached about 16%, the biospherians began to experience symptoms of high-altitude sickness (headaches, shortness of breath, trouble sleeping). Roy contacted four scientists to help figure whether the biospherians would adapt to declining levels of oxygen in the atmosphere. Dr. Harvey Meislin, our main consultant at the University of Arizona Medical School, Dr. Peter Hackett of the University of Alaska, Anchorage, and Dr. John West, of the University of California, San Diego, thought that humans could adapt to lowered oxygen levels corresponding to those found at altitudes of up to 18,000 feet above sea level. They also warned Roy that there would be an appreciable difference in individual adaptation, and that it might not be a straightforward process because the decline of oxygen we were experiencing was occurring over a long period of time. A fourth consultant, Dr. Igor Gamow, a professor of chemical engineering and sports physiology from the University of Colorado, predicted that we would not adapt. In his opinion it is not only the diminishing amount of oxygen in the air that is critical to acclimatization, but also the diminishing air pressure of high altitudes. For us, the air pressure remained the same during our two years.

During the year of decline, Roy closely monitored all of us. He sent blood samples to several different labs to track hemoglobin and red blood cell numbers. These are two factors which normally increase with adaptation to lowered oxygen — they allow the bloodstream to carry more oxygen to the cells of the body. Roy and Taber initiated a program to track our physiologic changes with a standard set of exercises. Every few weeks we all went to the analytical lab where a stationary bike was set up. We would do two minutes of bicycling, keeping the gauge between 600-700 calories per hour. Meanwhile a forefinger was inserted into a device which measured both heartbeat and the level of dissolved oxygen in the bloodstream. These were monitored for a couple of minutes before we started bicycling, and for four or five minutes afterwards until the heartbeat slowed down and stabilized. Taber also hooked up a tube to a gas chromatograph in the analytical lab. By having us expel most of the air in our lungs and then blow through the tube, he was able to measure the oxygen content in that last portion of lung capacity. By the time of the oxygen injection in mid-January 1993, our oxygen levels had reached about fourteen percent. This level corresponds to an elevation of 13,500 to 14,000 feet. Some degree of acclimatization (a small increase in red blood count) was measurable at this point. In other words, it was likely that we were slowly adapting, while we continued to show the symptoms associated with oxygen deprivation. The worst symptom was a decline in work capacity. This is difficult to measure, but it has been estimated that for every 1,000 feet above 5,000 feet, work capacity will decline by three percent. So by the time we injected oxygen into the system we may have suffered up to a thirty percent reduction in our work capacity. In fact, we had to scale down some of our operations in response. This was an additional factor in favor of importing oxygen into Biosphere 2.

For several months starting in the fall before the injection of the extra oxygen, Roy, Jane, Linda, and Taber received oxygen supplements through tubing that ran to their rooms. This oxygen was concentrated out of the Biosphere 2 atmosphere, it was not an addition of oxygen from the outside. This was done to relieve a symptom called Cheyne-Stokes respiration which would occur when they were sleeping. Cheyne-Stokes is triggered when the body senses that it is lacking oxygen. When a sleeping person with this disorder takes a breath it can end in a sudden gasp which would wake them up. Luckily the additional oxygen supply improved their sleep.

Roy alerted Mission Control in December 1992 that if he found nothing exceptionally interesting in the December blood analysis, he would recommend an oxygen injection to bring the oxygen levels back up to between eighteen and nineteen percent. He also recommended that sixteen percent be the cut-off point in the future because under that level the stress factor markedly increased. Several months later, Roy received a letter from the U.S. Navy requesting information on the experiment. The Navy was interested in applying this new knowledge to their submarine program — they want to be able to maintain submarine crews at lower oxygen levels because the lower oxygen minimizes the danger of fire.

Once the full amount of oxygen had been injected, all of our symptoms disappeared. Linda and Mark began to run joyfully around the perimeter of the domed west lung. Our friends on the outside watched us with wonder and curiosity as we all laughed and talked with each other. Just minutes before we couldn't have climbed a single flight of stairs without panting at the top. What an experience! First we endured a slow drop in oxygen with no corresponding change in air pressure. Then, suddenly, we received a large input of oxygen that gave us a huge lift. The eight of us will probably never take oxygen for granted again.

CHAPTER 7

The Wild Side

❧ *The care of wilderness*

Inside Biosphere 2, the wilderness cannot be taken for granted. On planet Earth most non-tribal people except for explorers and naturalists are only vaguely aware of the wilderness that sustains and nourishes life. Inside Biosphere 2, none of us can escape realizing the importance of the wilderness areas to our lives and the majesty of its operation.

SBV created the wilderness biomes from examples which exist in the equatorial regions of the planet. Teams composed of ecologists, entomologists, engineers, soil geologists, botanists, zoologists, microbiologists, and climatologists selected and combined species, communities, soils, rocks, and water flows to make small but richly diverse and authentic ecologies. These biomes require humans and a technical infrastructure to maintain the climate and certain other conditions necessary for their survival. They are not only beautiful, but essential for maintaining a breathable atmosphere. They also provided us or our animals with an important part of our food. In return, we maintain the conditions necessary for their complexity and diversity to flourish.

The biomes selected by SBV were the grand types of Biosphere 1: forest, grassland, desert, ocean, marsh, agriculture, and city. To maximize productivity of this basic range of biomes,

The wilderness encompasses an ocean, marsh, rainforest, savannah, and desert.

they were designed to be tropical. And from the tropics, they were designed to be those in monsoonal rainy climates. Thus they are the rainforest, the coral reef, the mangrove marsh, the coastal fog desert, the monsoonal savannah, and the double monsoonal agriculture. The wilderness areas were to be penetrated only by small trails just as tribal people use, or as a naturalist walks through the vegetation with minimum disturbance on off-trail areas. The wilderness areas were mainly to be managed by climate control based on their evolutionary experience. In cases of an aggressive species, the humans could interact as 'keystone predator'.

Biosphere 2 is a new kind of laboratory in which microcosms of the vast tropical regions of planet Earth can be studied to understand how they function and what role they play in the global cycling of specific elements and compounds. We can get deeper answers to such questions as: "What does a rainforest do for a biosphere?" or "Are all these different kinds of biomes necessary?" In addition, these tropical biomes are now among the most threatened ecosystems on Earth. Re-creating them in miniature and intensively studying them within a sealed system will provide valuable information about how we might restore damaged natural areas. Biosphere 2 may be the most important restoration ecology experiment ever undertaken.

Packing five different kinds of wilderness biomes in a space that is only about two hundred by five hundred feet constituted a unique challenge which attracted many of the outstanding wilderness experts in the world, including Drs. Richard Evans Schultes, Ghillean Prance, and Michael Balick. The different biomes must be separated from each other by special transitional regions, called ecotones. Between savannah and desert is a thorn scrub; between rainforest and ocean is a bamboo grove; and the fresh water marsh, salt water marsh, and ocean can be operated with separate water systems. Each of these ecotones demands special conditions, intermediate between the biomes.

The ecotones might have been taken over by an expanding biome if we couldn't keep them at different temperatures and rainfalls. Air handlers and air vents, sprinklers and pumps make us masters of the climate inside Biosphere 2, with one exception: we cannot control the sunlight. The tropical biomes we have replicated inside our biosphere are used to more intense sunlight, and we were eager to see how well they could adapt to the sun in the latitude of Arizona. But the most critical question was whether or not we had designed and built the biomes well enough that they themselves would tend to preserve their own diversity and integrity — with a little human help.

❧ *Humans as naturalists, humans as predators*

We had to be able to sense if an ecosystem or species was in trouble before it became too late to do anything about it. Gaie has spent years of her life wading through the muck of marshlands and diving among coastal reefs. Linda worked in deserts, Mark in deserts and savannahs, and all of us have made expeditions to rainforests. Our consultants were not only outstanding theoreticians, but at home in the wilderness biome they loved. These field experiences were essential because they alone could give a standard by which observers could assess the health and composition of our artificial ecosystem. We have a sense of how they should look, feel, and smell, and these observed sensations give the initial clue that something's up — that something needs further observations to understand. The skills of a naturalist were required by SBV for all biospherians to live and work in Biosphere 2, and our stay there helped hone our skills.

In addition to developing naturalist skills, the research teams selected some of the more complex and highly organized species that operated as 'bio-indicators' about the health of the overall system. These species would be the first to show symptoms if their eco-region was deteriorating. Sometimes, these are the natural keystone predators, the species that keeps other species of the food web in check, and whose continued existence shows that all the underlying links are still present. In the ocean for example, there are specific fish and corals which quickly show the first signs of trouble when the system is stressed.

We humans intentionally played an expanded version of the role of keystone predator for those parts of our small world. We were on call to intervene when required to cull out predators that had become too large or too numerous and so threatened the ecosystems which constitute their prey. We intervened when a plant species threatened to disrupt its eco-region or invade others. Learning when and how to intervene was a fascinating part of watching our biomes grow and mature. The mission rule was to keep interventions to a minimum.

Intelligent and judicious intervention by the biospherians assisted Biosphere 2's move toward self-regulation. It would have been impossible for this new world to operate without human participation except by severe diminution of species. First, we depend on man-made machines to assure the health of the system. Second, the wilderness ecosystems require individual and continuing attention. For example, the ocean and marsh required approximately four man-hours of attention a day. If this attention had not been given, the plants and animals could not have thrived. It was especially delightful to see how the systems immediately responded to our efforts — for example, when we installed the skimmers lowered the nitrates in the ocean.

Another major factor which the human contributes is the movement of biomass. In Earth's few true remaining wildernesses, this is done without human intervention: the massive power of organisms from microbiota to insects to animals consume, churn, tear apart, scatter, deposit, tunnel, displace, crush, and compost matter. Inside Biosphere 2, we could not wait for these processes to be performed because they had evolved over millions of years on Earth. We had to concentrate and accelerate the evolutionary process. Marine algae was moved out of the ocean at a rate of approximately eleven pounds a week. We collected compostable material from the agriculture area at a rate of one hundred pounds a day. We fed the animals sixty pounds of fodder a day, and pruning the wilderness and savannah grass produces about one hundred and fifty pounds of material a week. On top of this, we produced about 270 pounds of human food per week.

This process helps to maintain the elemental cycling which underlies the continuous operation of the Biosphere. It is a new role for people. In here, we alter or speed up ecological processes consciously to encourage environmentally beneficial results. For example, without herds of grazing animals like giraffes or deer, we humans have to harvest the savannah grasses before the rainy season begins, to encourage new growth. The edible parts are given to the livestock. The non-edible material is dried so it doesn't decompose and respire into our atmosphere in the form of carbon dioxide. At another time, when our atmospheric carbon dioxide levels permit, the cut grass is returned to enrich the savannah soils.

✌ *Deciding on the weather*

In Biosphere 2, you can actually *do* something about the weather. In fact, rather than weather predictions, we had weather requests. It was one of our most important jobs to keep our biomes happy about the weather. Humans programmed the computers to regulate the air and water temperatures, humidities, rainfall, and breezes while making sure pumps continued to make tides, waves, waterfalls, and the meandering flow of water in stream beds. And the observations and sensors give the feedback that enables the system to improve itself.

Each of the biomes has substantially different requirements. The rainforest receives moisture year round, and temperatures must stay between 55° and 95°. The savannah is active for a portion of the year, but during its dormant season it receives no rain, the drought condition allowing the tropical grasses to build up their root reserves. The thornscrub and desert areas also receive rain only part of the year. Natural coastal fog desert regions require hot summers and cool winters for optimal plant health.

The marsh is active year round. Here salinity requirements, water flow, and rainfall are critical. Since it has freshwater marsh as well as salt-water marsh ecosystems, the sections must be kept distinct from one another. The fresh marsh receives daily rain, while the rest of the marsh has sporadic rain. Once rain enters the salty water, it is not a usable water source for the terrestrial systems and must stay in the marsh. We aimed at raining just enough to make up for what evaporates, so Gaie frequently monitored water levels there. In the event of a possible flood (e.g. a rain-water pipe bursting, spewing water into the marsh), there is a machine that will desalinate the marsh water and send fresh water back into the rest of the wilderness water reservoirs.

The ocean is tropical, emulating coral reefs and lagoon areas found in the Caribbean. Here it is crucial that water temperatures vary little between the seasons, so our ocean is maintained between 76° and 80°F. The Energy Center supplies hot water in the winter to heat the ocean and cold water in the summer to cool it. The beach, with its coconut trees and other salt-tolerant sand ridge plants, is rained on several times a week, but aside from that, there are no rain storms in the ocean. Instead, water is condensed out of the atmosphere, recycled through a rain tank, and put back into the ocean to replace what is lost through evaporation.

The wind patterns provide the means to regulate temperatures in the different biomes. Even though there are no walls separating the rainforest from savannah or desert, the wind flowing from differently heated or cooled air handler units helps maintain the required air temperatures and humidities in these biomes. Winds also play an important role in pollination and air-borne ecological processes, not to mention the pleasure they provide to humans. But the maximum Biosphere 2 breeze was just five miles per hour — no squalls or hurricanes in here!

To create the climate for each biome, biospherians programmed control units which regulated the cooling or heating water that flows into the thirty air handler units or into the heat exchange system for the ocean, both of which are located below ground. The Energy Center supplies three sources of water for this purpose. The coldest is the chilled water supply which is generated from ammonia freezers at about 40°- 48° F. Next is the water from the evaporative water cooling towers at approximately 50°- 65° F, and then the hot water which is supplied from water warmed by the waste heat from the electric generators in the Energy Center at 180° F.

The water comes into Biosphere 2 in pipes which remain closed to the internal environment. These pipes enter through the Biosphere's walls carrying water that will never come in contact with air or water sources inside. Once the energy is exchanged in the air handlers or the ocean heat exchangers, an appropriately hot or cool air or water will be blown or pumped back out into the biomes. In the summer this generates cool breezes in the tunnels or warm spots in the cold ocean.

Hot water is only needed in the winter time to ensure that temperatures remain tropical and don't fall below minimum levels. By far our greatest energy needs were in the warmer months when we had to use cool and chilled water to prevent temperatures from rising past what life in our biomes can tolerate. Without cooling water in the summer, the Biosphere could rise to above 150 degrees in just a few hours. That's why we have backup generators and a second set of closed loop water piping between Biosphere 2 and the Energy Center in case the first set of lines should spring a leak.

We not only programmed our automatic devices on when and how long to rain, but we chose the quality of the water that was used. A continual challenge in our water recycling was managing water quality, the amount of nutrients and salts it contains. The rain that falls in each biome produces some water that drains through the soil. This water, salty and high in nutrients, is called soil leachate, and it is collected in the basement and stored in tanks to be reused. Therefore, a process of testing the water commences in order to determine if it can be directly rained back onto the wilderness. Often the collected leachate water needs to be mixed with condensate water (a very pure water condensed out of the atmosphere) in order to dilute the salt content.

The terrestrial biomes do not like salt. The ocean and the marsh, on the other hand, will handle the salts but cannot tolerate high levels of nutrients. We had to keep a careful eye out for the build up of either salts or nutrients with the long-term goal of assisting the cycling of certain elements in our biomes. Occasionally the agriculture system will also be involved if it runs out of condensate generated from the IAB and needs some condensate water from the wilderness. At times, we had intense negotiations over allocation of condensate water, each area manager arguing his or her case for receiving more.

The 'rain' comes mostly from sprinkler heads mounted overhead in the spaceframe. In some special spots (where we don't want to get walkways or ventilation gratings wet), there are also ground sprinklers with directional heads. As trees grew in the rainforest and started to interfere with an even flow of falling rain, we had to lay down soaker hoses to ensure that every area received adequate moisture. It is extremely important to provide the right amount of rainwater.

Some of the most delightful contrasts came from watching our biomes go through their seasonal changes, as outside, the high desert mountain ecology of southern Arizona goes through a quite different series of seasonal changes. It's often quite strange: we can be walking through a light rain shower in the rainforest while outside it may be winter and snow covered, or sizzling at 105 degrees under the dry desert sun. Conversely, we may be working in the dry and dormant Biosphere 2 desert, while the sky is black with rain clouds and the landscape outside is inundated with a monsoonal downpour. At such times, with the raindrops beating on our glass roof, we could almost taste the outside rain.

In some areas where slopes are steep, like in the east gingerbelt of the rainforest and the desert bajada, we had to be careful about soil erosion. After closure, we spent many hours replanting ground-cover vines and plants in the rainforest where the footsteps of construction crews and guest tours before closure prevented them from becoming established.

Even now, with such 'bonsai' biomes, we must be very careful about avoiding excessive trampling. Narrow trailways were established in each biome for biospherians to follow. No one ever veered off the track unless management or research required it.

One of the ecosystems in the rainforest requires year around fog. This is the cloud forest area of the rainforest modeled after the Amazon cloud forest 'tepuis' where the plants are naturally covered in almost permanent low-lying clouds. To produce this fog, we used artificial misters. Every five minutes, spray nozzles eject a fine mist around the top of the rainforest mountain bowl in which are planted ferns and vines around several small ponds. When the misters are on, visibility can fall from six to twelve inches — a man-made cloud.

There are two streams in Biosphere 2: one which meanders

in wide loops through the varzea (flood plain) of the rainforest, and one bordered with African acacia trees to create the shady 'gallery forest' of the upper savannah. Biospherians regulated the flow in these streams, periodically cleaning out any vegetation that may choke the passageway and ensuring that their recirculating systems are operating. The varzea stream water is continually pumped back up to the interconnected ponds at the top of the cloud forest. From there, water overflows and creates the waterfall that tumbles some twenty-five feet to the pleasantly cold and deep 'Tiger Pond' at its base and thence back into the varzea stream. The savannah stream, when it reaches its terminus at the southern border of the upper savannah, is automatically pumped back to its head, where a small waterfall cascades down to commence the stream run.

✣ *Helping the rainforest grow*

The rainforest was one of the most challenging of the terrestrial ecosystems we had to build from scratch. Arizona is far from the tropics where day lengths vary only slightly from season to season, and to start with, we had to protect the light-sensitive plants from the blazing desert sun overhead. Dr. Ghillean Prance, then Vice-President of Research at New York Botanical Garden, and now Director of the renowned Royal Botanic Gardens at Kew near London, is one of the world's greatest experts on the Amazon rainforest. He came up with a strategy that he thought would not only be successful for Biosphere 2, but would also be a first trial for a method that would be useful in reversing the devastation of the Amazon itself. The first element was a fast-growing selection of trees that would quickly form a canopy to shade and protect the slower growing rainforest trees that would constitute our mature system. While those trees would form a protective shield against overhead sun, a thick gingerbelt planted around the periphery of the rainforest would screen out the harsh sun on the sides. The gingerbelt is populated by true rainforest species, but ones which tolerate strong sunlight and are fast growing: banana, canna, heliconia, and ginger.

When the biospherians worked with a team from the Yale School of Forestry in 1990 and 1991 to measure the initial size of all the plants before closure, the tallest trees were under fifteen feet. It required considerable imagination to picture this scrubby scene as the site of a future noble rainforest. The bamboos of the bamboo belt, which were planted along the beach cliff to protect the rainforest from salt particles blown from the ocean, were just a couple of feet tall. The varzea was still exposed to harsh overhead sun and the soil along the stream banks was held in place by burlap as a stop-gap measure until the plants sent out their roots to hold the soil in place.

In any event, the rainforest thrived during its first two years. By the first anniversary, biospherians had logged many hours cutting back trees that were touching the spaceframes fifty feet above the ground. Some of our first-canopy trees, Leuceana, Cecropia, and Ceiba, had grown more than twenty feet the first year. Almost overnight, they created a rainforest with towering trees protectively shading the lowland rainforest. The bamboo belt had clumps that were over fifteen feet tall (now, they are approaching thirty feet). In the varzea, modeled on a 'black river' area of the Amazon, the small Phytolaca trees we planted grew both up and out. Three trees now clothe the varzea with heavy shade, with tree trunk diameters that you can barely stretch your arms around.

The gingerbelt also worked as designed. It became so dense that it completely blocked out views of the Mission Control building a few hundred feet to the east. Because it's fast to regrow, it plays a valuable role in our management of wintertime carbon dioxide. We prune the gingerbelt back and store the biomass dry in our technical basements so it doesn't decompose and release carbon dioxide. Meanwhile, since the gingerbelt rapidly regrows, the plants very effectively take up carbon dioxide. The gingerbelt can be cut two or three times during the fall and winter months if required, and in the summertime, this dried biomass is returned to the soil to release needed nutrients.

The rapid growth of the first canopy trees created a problem of its own. Because the trees grew so fast and because there is less overall light inside Biosphere 2 (due to light reduction caused by the glass and spaceframe) as compared to the natural Amazonian levels, their limbs became weak as they stretched for the light. Their top growth became disproportionate to their root growth. That, combined with the lack of strong winds, means the trees lack 'stress wood' — the stronger wood that allows trees to handle strong wind impacts. So some of the tall rainforest trees have started to lean and a few have toppled over. In the upper savannah, several acacia trees, which have also grown tremendously, have arched right over or had their top branches break.

Because of this, one of the periodic wilderness jobs is to decide whether to straighten the trees with braces, or take weight off them by pruning back some of their top branches, thus allowing the roots to catch up in better anchoring them. The former decision calls for biospherian acrobatics. One of us climbs high in the spaceframe to attach a rope from the roof to the tree and then it is winched more upright. In some cases, we leave the rope as a long-term assist; in other cases long wooden braces are cut and attached on three sides of the tree to prevent any further leaning.

✤ Tall and weedy tales

One of the dire forecasts that some scientists made before we began the experiment was that we wouldn't be able to maintain separate biomes because weeds from one would invade others; and each would be swamped by its own rankest growers. While that hasn't proved true, a few plants have required considerable effort to control.

During the first six months of our stay in Biosphere 2, we all admired the morning glory on our weekend strolls through the biomes. Covered with blue flowers, the morning glories draped twenty feet down the sides of the cloud forest. The modest growth of the domesticated flowers you may know from your garden gives little indication of what prodigious feats a morning glory is capable of given an environment it loves. Our jungle morning glories were not content to cover the sides of the cloud forest. Soon they had spread a dense cover of their leaves over trees on every side of the cloud forest. They even forded the Tiger Pond (which stands at the base of the waterfall), clogging the water with hundreds of pounds of vines and roots. Then they marched south along the ground and into the trees of the varzea and clear over towards the beach as they fell over the cliff.

Wherever their long vines dropped on damp soil, they put out roots.

The morning glory probably sneaked into the rainforest in the form of a handful of seeds. It must have produced several tons of biomass during its reign as the King Kong of the cloud forest. At first, we welcomed the fast growth and its shade, also thinking of all the carbon dioxide the vines were soaking up. In addition, periodically cutting them back provided delicious fodder for the goats. But as time went on, our strategy of fast-growing trees succeeded well in providing shade for the slower-growing trees. The dense covering of morning glory leaves was too much. Tree branches sagged under the weight of the invaders, and understory light levels fell disastrously, even for rainforest species. As keystone predators, we decided that morning glory was not going to be part of our rainforest since it would take far too much time to keep under control. It had to be weeded out.

By the first summer we had started waging war against the morning glories. But we'd let them go for too long. Their vines were everywhere on the forest floor — a thick maze which tapped into the soil for hundreds of square feet from where they climbed down from a nearby escarpment of the cloud forest. It was no longer possible to tell where any of it started or ended. And any bit of the vine left uncut or uncollected would regenerate itself and start relentlessly growing. When we halted the morning glory advance they were about to cascade from the varzea and start the invasion of the upper savannah to the south!

In the final year we waged a systematic campaign to find and eliminate every last bit of morning glory vine before the end of the two-year mission, so that it would not be a burden on the next crew. Taber and Linda went into the spaceframes to liberate the trees from the vines (working before breakfast before the air up there became too hot). Mark did much of the ground patrol work, cutting and bundling up morning glory vine like so many rolls of electric line.

There have only been a few other plants which, like morning glory, have proved to be serious weeds, or required much time for cutting back. One of them, curiously enough, is the passion fruit vine. Unlike morning glory, passionfruit was a planned member of our biomes. In the rainforest, we planted several varieties of this tropical plant to provide genetic diversity and food. In the savannah, it was planted under some of our African acacia trees in the gallery forest that fringes the savannah stream. We were rewarded with a rain of passionfruit during the summer and fall months.

Passionfruit makes a tasty ice cream flavoring and lends a special zing to our fruit juice mixes for feast drinks. But, somewhat like the morning glory, the passionfruit vine demonstrates how prodigious plant growth can be given the right conditions. A vine on the rampage invades new territory and smothers whatever is in its path. Our savannah passion vines have covered whole areas of the savannah in a grape arbor-like web overhead, bending tree branches with their weight. In the east rainforest, they have ascended from the top of the gingerbelt plants and created an impenetrable mass of leaves which covers the higher spaceframe glass.

In the summer, we let the vines grow when there was abundant light for the other plants. This gave extra shade to sensitive trees underneath. In winter, when light is very limited and our rainforest trees go half-dormant, we pruned back the passion vine to let more light reach the under story plants.

In the desert and lower thornscrub, the major invader was Bermuda grass. We suspect it came in with some of the local soils, but with the year-round tropical climate of Biosphere 2, it's far more productive than it is in the deserts of Arizona. Compared to the morning glory, however, the grass problem is far less serious. For one, the desert and thornscrub areas go dormant for part of the year and, without water, the Bermuda grass can't grow. It's been more of a problem in the sand dune area of the desert. There are more empty spaces there, and a variety of grasses now grow in between the sand dune bushes and the creeping devil cactus.

By our second year, the ocean had become an underwater garden that required extensive weeding. Gaie would dive up to four times a week to pick the colorful, large algae from around the perimeter of corals and from the ocean floor where it accumulated in floating islands. In addition, there began to grow a marine chlorella algae all along the rocky surface which became a home to small crustaceans called tenaids. Fresh water chlorella algae is notorious for being one of the fastest growing algae species and had been used by the Russians for atmosphere regeneration in small closed systems. So it's not so surprising that a marine species of chlorella found its way into the system and took advantage of the high carbon dioxide and available nutrients to grow everywhere.

Part of the maturing process of our ecosystems will be their ability to resist the invasion of foreign weeds. We probably did more weeding than future crews will do, because the biomes had not yet evolved a hardy, ecological community. We've already begun to see in these two years that, as trees get taller and their foliage shades the ground more, it becomes more difficult for weed invaders to grow below them. In the upper thorn scrub, the grass problem is already far less serious than it was at the beginning of the closure since the thornscrub trees have doubled in height. In this sense, the biospherian keystone predators just lent a helping hand in these early years while the biomes organized themselves.

❧ The desert shows a mind of its own

Life asserts itself in surprising ways, and this is particularly true with the desert. When SBV designed it, a basic problem had to be solved: how to find a desert that could live with the high humidity which could not be avoided with an ocean, marsh, and rainforest under the same roof. The research team found the solution in the 'fog deserts' of coastal regions. For part of the year, heavy fogs rise from the ocean currents and cover the vegetation, supplying some of its water needs and reducing losses due to transpiration. These deserts have adapted to high humidity, though rain is still extremely sparse and sporadic as in all desert regions.

Fairly close to Arizona is one of the most striking fog deserts — that of Baja California. Collections from the plant life there form the majority of the Biosphere 2 desert, although some plants from Israeli, African, and South American coastal deserts were also included. The semi-scrub plants, desert bushes, and annuals (whose seeds were collected from Baja) grew luxuriantly in our conditions, enjoying the longer rainy season we gave them the first year in our attempts to keep carbon dioxide levels low.

In places where condensation off the glass added extra moisture and extended the growing season before the soil dried out, they did especially well. The salt bush family (Atriplex), formed a bonsai forest on the slope down to the salt playa. It was only in limited areas of the desert that the boojums, yuccas, and columnar cardon cactus (a fog desert relative of the Sahuaro cactus which grows on the mountain slopes outside Biosphere 2) dominated, giving the kind of cactus and succulent desert originally pictured.

The bulk of the desert developed into what ecologists call a coastal sage scrub desert environment. Instead of trying to force it to stay in the original conception, we went along with its own tendencies. We included a good diversity of plants in our original introductions so there is still a good mix of species. We will introduce yet more diversity during the transition, enriching the kind of ecosystem that is emerging on its own.

The two thornscrub ecosystems have developed as anticipated. They were put in as transition zones between the savannah and the desert, for that is how they are frequently found in the tropics. The upper thornscrub is primarily based on communities found in Sonora, Mexico, and the lower thornscrub is dominated by Madagascar/South African thornscrub species. The night-blooming cereus cactus from South America is a special plant in the upper thornscrub. Its beautiful white blossoms appear for just one or two nights at the beginning of the rainy season.

In both desert and thorn scrub, the biospherian weathermakers had to pay careful attention to the plants to gauge when to start the rains and when to end them. This is far more important than it is in the savannah, which has adapted to a variety of rainfall seasons and more frequent out-of-season storms. The desert and thornscrub species will not withstand rainy seasons much different from those of their natural habitats. Since many vectors are somewhat different inside Biosphere 2 (such as day lengths, temperatures, the mix of plants) we watch the desert and thornscrub carefully to pick up signs that the plants are ready for rain. Certain indicator plants are frequently checked to see if their leaf buds are swelling or if flowers have appeared on those which bloom just at the beginning of the monsoon season in their native habitats. When the plants signal yes, the irrigation controls which program the rain are set. The word is passed on: "On Thursday at 5 PM, the north desert will receive its first rain;" or "Saturday morning right before breakfast, the upper thornscrub will get an initial rain lasting forty-five minutes."

We also made frequent checks of the desert and thornscrub to gauge when the rains should be stopped. While the desert plants can withstand a short rainy season or a drought year, they may have a harder time with too much rain. In these initial years of operation, we wanted to keep them active as long as possible, so we checked key species for signs of root or stem rot that might indicate we were keeping them active too long. When the rains are stopped, the biome dries out and becomes inactive again.

❧ Home on the range, where the tortoises graze

The savannah is more flexible than the desert. It contains a wide spectrum of grasses that were included from all the world's savannahs, — Africa, Australia, and South America. The challenge in this biome was to replicate the feel of the open horizons of a savannah in barely half an acre. The savannah is long and narrow, bounded by the cliffs that separate it from the ocean and marsh to the east. The whole structure slopes to the south, where the desert lies at the lowest elevation in Biosphere 2. This allows hot, moist air to rise and flow uphill towards the rainforest at the north end of the wilderness biomes. The savannah is also divided into an upper savannah where the stream and gallery forest of acacias and passionfruit are planted, and the lower savannah which forms a small ocean of grasses.

Savannahs are tough environments, exposed to seasonal flooding and long dry seasons, with far more scorching temperatures than the rainforests. The grasses which dominate them are a recent evolutionary addition, and they grow continuously to survive the teeth of grazing animals. In fact, most savannah ecologists agree that savannahs need disturbance to stay healthy. Grazing stimulates the grasses to extend themselves with underground rhizomes, and to put up fresh green growth. Most grasses have adapted to the heavy hooves of range animals and to periodic fires which clear out the old growth. But fires are forbidden in Biosphere 2 (because it would pollute the atmosphere) and our savannah is far too small to even think about introducing any of the large grazing animals which form the herds of the African savannahs.

At one point, we contemplated the idea of occasionally taking our goats out on leashes to graze. This was impractical for logistical reasons, so we looked for wild grazers that could be supported. We chose leopard tortoises from Africa which were introduced as a herd of three before closure. But we had few illusions that these twelve-inch-long animals could make much of an impact on our grasses!

Put a sickle in your hand, and start to hack your way through the savannah, and its size suddenly becomes magically magnified. We'd experienced that the first fall. The lower savannah, grown into a crowded tangle of four-to-eight-foot high grasses, was cut step by step by advancing armies of biospherians. At the end of several weeks, our little plot seemed like vast grasslands. When we shifted to the upper savannah, the billabongs also commanded respect. These areas are named for the low-lying semi-lakes of Australia, which are actually running waters during the wet season, but which gradually shrink in depth and sometimes dry out completely during the off-season. The Biosphere 2 billabongs are about twenty-five feet long and twelve feet wide, and seasonally fill with the overflow from the savannah stream. But try to cut your way through them and you discover what kind of growth some grasses are capable of! We pulled out some that were twenty-five feet long.

The African acacia trees have also made themselves at home. When they were planted, just a year before closure, they were thin and straggly. Now several of these trees are approaching the roof, thirty-five feet overhead. Festooned with passionfruit vine, and other vines from the savannah stream that have hitched a ride on their branches, they are beginning to give the landscape the special look that acacias give savannahs worldwide. While the grasses learned to live with the grazers by outgrowing their teeth, these African acacias employ a counter-strategy: nasty thorns to convince the wildebeest and giraffe that acacia leaves are no free lunch. We traveled the savannah in thorn-proof shoes, pruning shears at the ready to lop off dangling branches. In this respect, the acacias are like the thornscrub areas which divide savannah from desert.

❦ *The water worlds*

Marshes are known for their high levels of organic sediments, high-nutrient water sources, and richly diverse marine plant communities. Coral reefs are entirely the opposite, with porous, sandy sediments, low nutrients, and only a few underwater plant species. Key species in the marsh are the mangroves: white, black, and red mangrove trees reaching into the salty water, masters of bridging the terrestrial and salt water realms. The peculiar branching root structure of the red mangroves, gray-white oyster beds that rise out of the water, and dense green foliage hanging suspended over dark green and brown marsh water are the characteristic traits of the mangrove marsh ecosystem, forming a network of entangled roots. The coral reef, on the other hand, is the secret realm of the ocean. If the observer gazes on top of the water, there is nothing to see except coral heads here and there breaking the surface of the water. But under the sea is the colorful, dazzling, and magical reef world with caverns, currents, fish schools, and large sea animals that cruise in and disappear as quickly as they appear.

Reefs are notoriously difficult to maintain. We made a big leap from what was known about aquarium exhibits to build the 900,000 gallon marsh and ocean. For a start, these tropical, largely equatorial marine communities are nestled in the foothills of desert mountains, approximately thirty-three degrees north of the equator, 3,900 feet above sea level, thousands of miles from the nearest tropical ocean, and controlled by the regimes of a temperate light pattern changing markedly in intensity and duration between seasons.

The marine life is entirely dependent on supportive

technology. Waves are generated from an approximately fifty-foot-long vacuum system that sucks approximately ten thousand gallons of water into a ten-foot-high chamber and then releases it at the south end of the ocean to create a one-foot-high wave. Currents and flows are made by more than twenty pumps. In the marsh there are pumps for each of the six different ecosystems to provide currents within that particular water body.

Initially, we operated with a tidal system to create tides for the marsh and ocean. This required pumps, controls and levers to move approximately 6,000 gallons of ocean water into the red mangrove marsh which flows slowly up to the freshwater marsh section which is at a higher elevation. Once a high tide is reached, the water then flows back to the ocean controlled by computer until the low tide level is reached. Because the ocean required extremely different water quality from that found in the marsh, we shut the tidal system off in November 1991 and from then on operated the ocean and marsh as separate water systems.

Once constructed, monitoring of the chemical and physical properties of the marine systems is critical. This is where most of the initial biospherian time was spent in assisting marsh and ocean, as opposed to the time spent in the wilderness areas managing the excessive growth of weeds. We collected water samples weekly for nutrient analysis, and for measurement of pH (measure of acidity), dissolved oxygen, and other parameters and made twice daily checks of all the technical systems to be sure they were still operating correctly. Monitoring the water chemistry of marsh and ocean provided essential information about the health of the systems.

The mechanical checks were critical because the water

systems must have motion and flows to support higher life forms. If water flows stop for more than twelve hours, the system goes on a red alert. Several times the wave machine caused problems, both from its mechanical and computer controls. No matter what time of the day or night the system went down, Laser and Gaie would work as long as necessary to get the system back up and running.

Our primary concern for the ocean was whether its complexity could survive the dramatic fluxing of carbon dioxide. As the concentration of CO_2 increases in the atmosphere, it diffuses into the ocean water and this lowers the pH of the water (making it more acidic). Coral reefs are found in waters where pH ranges from 8.0-8.4. With the high levels of carbon dioxide that were predicted in Biosphere 2, we needed a method to 'buffer' the ocean in order to keep the CO_2 from lowering its pH. Therefore, we stock piled fifty-pound bags of carbonates and bicarbonates of which an approximate 4,500 pounds were added throughout the two years to counteract the rises of carbon dioxide. This chemical effectively ties up the CO_2 as it dissolves into the water and increases the ability of the water to withstand an influx of CO_2. Even though this helped keep levels of pH from plummeting, the levels reached were still low in comparison to natural reefs.

Despite this, the reef did extraordinarily well. We had no prior information about what would happen should pH reach the levels it did. The ocean proved after two years to be far more adaptable than we had thought possible.

We had a number of jobs to do besides weeding the ocean. For example, there was an outbreak of fire worms, a red-orange, prickly worm that eats soft corals and, in particular, anemones.

They were not intentionally introduced into the system, but slipped in with the reef rock. They quickly grew into a serious problem. For several months shortly after closure, it was not uncommon to hear tour guides on the radio reporting a fire worm sighting through the ocean window to alert Gaie so she could quickly don her diving gear and remove the worm. Several hundred were removed, and although there are still some small ones found from time to time, they are no longer a problem.

Another species that required culling were Spanish and striped lobsters. Three species of lobsters had been introduced into the system as small adults. By the end of the first year, grown much bigger, they began decimating the snail population. Snails are key grazers of small algae, and without them, the algae would smother our corals. After some lengthy discussions about the matter, the research team decided to remove two of the species (leaving the third, the spotted lobster) in order to give the snail population a chance. The lobsters became a delicious meal, sautéed with mixed vegetables.

The parrotfish and squirrelfish continued to be a problem. Individuals in both species have grown considerably. The parrotfish, the largest fish in our ocean (some have grown to about a foot and half long) from time to time take bites out of the corals. The squirrelfish patrols for new fish babies, consuming them before they can grow to adulthood. Laser tried to hunt the largest of the offenders but to no avail. The reef is quite porous and it provides ample opportunity for the fish to hide. Techniques used commonly on Earth to catch fish in open water, such as harpoons and nets, don't work in this coral garden designed with complex econiches to hide in. This will undoubtedly be an area where fine-tuning in management skills or changing the choice of species to complete food webs is required.

The marsh is the easiest of systems to maintain. In general the marsh has flourished. The mangroves are now soaring twenty feet above the water; when they were introduced three years ago as young trees their a maximum height was about five feet. Aside from the increased biomass, the marsh may prove to be an essential component in closed systems because it is extremely resilient. It has the capacity to receive excess salts or nutrients in its sediments without causing damage to the other life forms.

❧ *Passing on the torch*

Our biomes have gone a long way towards growing up in the two years that we have been their allies and occasional caretakers. But they have a long life ahead of them, probably decades of continued growth to maturity. Though our biomes are modeled on the tropical regions of Earth, they are becoming unique and distinctive ecosystems, adapting to environmental conditions unlike those found outside Biosphere 2.

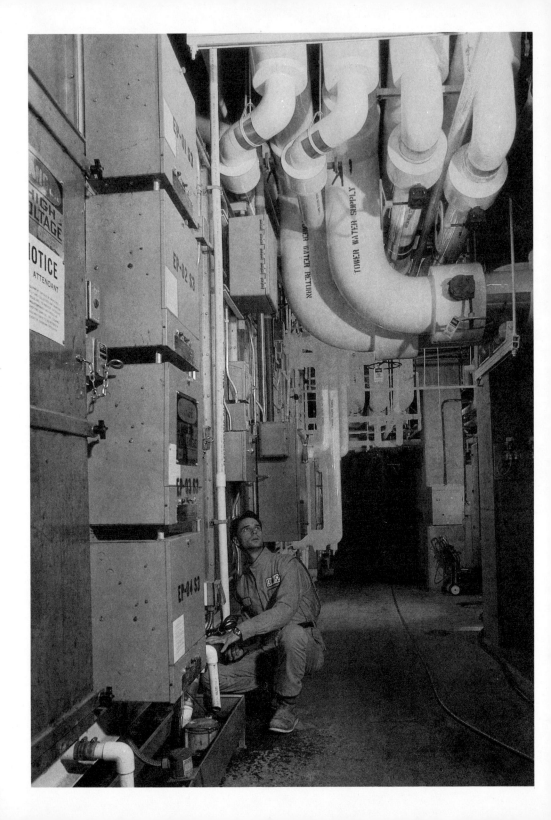

CHAPTER 8

The Technosphere

Biosphere 2 has two faces, one above ground and one below. Above ground are the abundant, varied life forms of the wilderness areas, while the underground environment is like a battleship, full of humming and whirring technology that supports the wilderness above. In these basements are some one hundred twenty pumps, forty-five air handlers, several miles of electrical wire and pipes, water storage tanks, computer controllers, filters, an algae-based nutrient removal system for the ocean and marsh, systems for rainfall irrigation, heating and cooling exchangers, desalination systems, a chemical recycler for atmospheric carbon dioxide, diving equipment, composting equipment, and other assorted machines. In addition to this, a workshop divided into six areas, (mechanical, plumbing, electrical/electronics, woodwork, rigging, and spare parts storage) was designed to let us handle any maintenance or repairs without having to rely on outside assistance. A complex network of computers, sensors, videos, and communications gear runs throughout the Biosphere giving us real-time status reports on the technical systems. In addition, a computer-accessed system of vibrational analysis tracks many of the key technical components of Biosphere 2's systems and can give us advance warning of impending breakdowns.

Hundreds of pipes, pumps, and other mechanical devices help maintain the ecological environments inside Biosphere 2.

The smooth operations of the technical systems have been an extraordinary success. This is a tribute to the care and workmanship that went into Biosphere 2's design and construction. Laser, manager of the technical systems, has been able to work with biospherians to accomplish all maintenance and repair operations using less crew time than we had anticipated. Only eight percent of all biospherian work time is required for the repair, maintenance, and upgrade of technical systems. This comes out to less than thirty hours per week — under four hours a week for each of us. In addition, he has upgraded some technical systems making them easier to operate. He has also fabricated entirely new systems that we didn't realize we needed until after closure.

During the seven years it took to build Biosphere 2, Laser served as head of Quality Control. He was involved in every phase of the development of our major systems from blueprints to installation. He knew he was going to be the one that would have to maintain all of these systems during the first two-year mission. This ensured an active dialogue between construction and maintenance personnel during the critical design period. Usually maintenance men are specialists. They don't get involved with a system until after it is complete and then they discover the problems of actually operating it. In the case of Biosphere 2, Laser helped ensure that ease of maintenance was an integral part of the design. This fact is significant and may have a great deal to do with why our technical systems have performed so well. One of the key design criteria was the choice of lubricants, oils, and greases. We could not afford to use toxic chemicals that are not recyclable, so all these substances had to be what is called 'food grade'. After considerable research SBV found what we needed.

One of the main ideas behind the maintenance system for Mission One was to simulate as nearly as possible a two-year voyage to Mars in which the crew would have to rely on what was included inside the system. We planned to carry everything we might need. Back-up pumps and spare parts were included as well as the tools and machinery to fabricate virtually any piece of equipment. Laser was key to this strategy since he had most of the skills required for such a daunting task, and SBV sent him to intensive schools to acquire the rest.

For the first year, we operated with nothing coming in from the outside. In the second year our Scientific Advisory Committee recommended a limited amount of export and import to enhance the research program. We then allowed the import of spare parts if they saved substantial biospherian labor and the export of malfunctioning equipment for repair such as electronic or computer components needed for our scientific research program. The original idea of total self-sufficiency was very useful, since it encouraged the biospherians to troubleshoot their own systems and not rely in any way on outside specialists to fix something that was broken. This is where hands-on experience is invaluable. The person who 'can do' is vital to the operation of a closed system, rather than the textbook theorist who has never gotten his hands dirty. The ability to improvise and to 'jury-rig' equipment when necessary was a crucial aspect of living in a closed system.

❧ *Making the Biosphere air-tight*

The significance of the experiment lies in the ability of Biosphere 2 to be virtually airtight. From the beginning of the project we'd identified this as the number one technical engineering challenge. It was the task of Bill Dempster, our head of Systems Engineering, to reduce the rate at which air leaked out of or into the Biosphere, a task he worked at right up to closure and all through the first experiment.

Bill was notorious in the pre-closure months for spending long and tireless hours in the underground tunnels that surround the Biosphere. He wore headphones to magnify the hiss of air caused by tiny leaks along the stainless steel seams and joints which he intended to seal before closure, and he carried a telephone to transmit the leak's exact location to his assistants. It was mind boggling that he had the patience to do this minute detection work on a structure that was so enormous!

Bill wasn't the only one looking for leaks. There were also 'spaceframe men' with soap-bubble machines. Over twenty miles of glass seals were covered with soap bubbles (just like you use to find holes in automobile tires) as the first step in tracking down air leaks. Over thirty leaks were discovered in this way. To use the soap bubble method, the Biosphere had to have a slightly higher air pressure than the outside. This would force air out of the holes and make the soap bubble. Bill figured out that the best way to create such a pressure was by using two 'lungs' which he had designed for Biosphere 2.

In the lungs there are huge rubber membranes housed under beautiful white geodesic domes which are connected to Biosphere 2 by underground tunnels. As the air temperature rises during the day, the atmosphere inside the sealed structure expands and rushes through the tunnels where the lung membranes rise to accommodate it; at night, as temperatures fall in Biosphere 2, the atmosphere contracts and the lung membranes deflate.

This lung design enabled us to manage the air pressure and build the vast roofs of spaceframe and glass. Without the lungs, we would have had to reinforce our structure, or it might have faced the danger of exploding or imploding!

By using the lungs to exert a slightly higher pressure inside the Biosphere than outside, we were able to calculate the amount of air leaving the system by tracking the height of the lungs and comparing that with our calculation of what the height should have been without a leak.

In December 1991 we ceased creating a positive pressure to force air out in order to find the holes. Bill then relied on the concentration of two trace gases to show how much air was getting out. He introduced a known quantity of two biologically inactive gases, helium and sulfur hexafluoride (SF_6). In addition, the researchers from Columbia's Lamont-Dougherty Lab introduced krypton as an independent check on leak rates. Since these three gases are found in extremely small quantities in Earth's atmosphere, and we knew exactly how much was released into Biosphere 2's atmosphere, their decline over time gave us our leak rate.

The unprecedented low leak rate of Biosphere 2, under ten percent a year, is due to Bill's diligence and the superb efforts to develop new sealing techniques. To appreciate how airtight Biosphere 2 is (given its three-acre size, twenty miles of joints covering 6,600 glass panels, and its vast stainless steel underground liner), a one percent annual leak rate would be caused by microscopic holes altogether amounting in size to your pinkie fingernail! In contrast, a typical twelve-foot by nine-foot by eight-foot-high office at the EPA minimum standard of fifteen cubic feet of air exchange a minute has an annual air exchange of 100,000 percent which is 20,000 times greater than Biosphere 2. The Russians with their Bios-3 closed system (the previously most tightly sealed structure in the world), had a leak rate of about fifty percent per year. NASA, at its Breadboard facility at Kennedy Space Center, had converted a Project Mercury pressure chamber into a closed system for plant growth for space. They got their leak rate as low as three to five percent *per day* (more than 1,000 percent per year) and stopped there.

A living system, however tightly sealed, must still permit a sufficient flow of energy and information. Earth is a closed system like Biosphere 2, yet it remains open to energy and information from the outside. Earth receives energy continuously from the sun, which permits life to flourish here. If the Earth were denied a source of energy, it would shortly entropize, drop to a level of energy that could not support life, and become a chemical equilibrium like the dead surfaces of the moon and Venus. For Biosphere 2, energy is supplied in the form of sunlight and electricity. The electricity produces hot and cold water for heating or cooling and is supplied by generators (powered by natural gas) and other equipment in an Energy Center just a few hundred yards away. Biospherians send and receive information through computers, faxes, telephones, and video.

The Earth is almost, but not totally, materially closed. In addition to the loss of hydrogen and other light gases from our upper atmosphere, the Earth also receives over a hundred thousand tons of cosmic debris (meteorites, etc.) a year from space. Biosphere 2, through leakage and the export and import of research samples and tools has a ten percent exchange rate with the Earth's atmosphere per year. These imports or exports of scientific materials have a negligible mass as compared to the mass of the entire system, so these items are essentially an information exchange. Biosphere 2, like Earth, remains a materially closed system and will remain so when we have left.

❧ *Breakdowns and repairs*

There have been a few problems with the numerous pieces of equipment inside Biosphere 2. For example, one type of PVC piping that was installed didn't have a high enough PSI (pounds per square inch). It frequently failed in our wilderness water systems due to the high water pressure, so Laser went around and systematically replaced that type of pipe. Similarly, a few systems were made unnecessarily complex, so we simplified them. This was especially true of some of the computer programs for the technical systems. When Laser and the rest of us started using them it became clear they were not as user-friendly as they should have been. We appreciate the old engineer's motto: "It takes a genius to make a system simple. Any fool can make it complicated!"

A lot of the technical work involved the fabrication of new systems. In the ocean, we started the two-year experiment relying on a nutrient removal system called an algae scrubber to clean the ocean and marsh water quality. There were sixty of these scrubbers, each composed of a four-by-eight-foot fiber-glass box in which two plastic mesh four-by-four-foot screens were placed to facilitate the growth of a community of some forty types of marine algae. Water is pumped from the ocean and marsh and dumped across the screens as if from 'waves' via a plastic bucket which fills and splashes over when full. These algae scrubbers simulate the action of natural reef algae, which keep water nutrients low by using them for rapid growth. Every ten days, we had to scrape off the accumulated algae to keep the scrubbers usable.

We discovered that this system, in addition to requiring up to twelve hours per week of biospherian labor, was inefficient and incapable of doing the job required. So, six months into the experiment, Laser and Gaie decided to employ a different type of nutrient removal device called a 'protein skimmer'. Skimmers are composed of a fiberglass pipe through which air bubbles. As the bubbles rise against a stream of water, dissolved tannins, acids, and nutrients accumulate on the bubbles and get 'bubbled out.' This is analogous to the foam that accumulates at the edge of a beach from waves lapping along the shoreline. With the advice of aquarist Julian Sprung (Two Little Fishes) and the successful trial run of one skimmer installed by Keaton Fisheries before our entry, we designed and built seven new skimmers for the ocean.

Wisely, Laser had stockpiled lots of miscellaneous materials inside Biosphere 2 on the off-chance that we would have to fabricate some entirely new system. So he and Gaie took stock of the inventory of pipes, pumps, air compressors, air stones, hoses, and other miscellaneous parts stored inside the Biosphere and found what they needed to construct and install the skimmers. Immediately after they came on line the ocean water began to show signs of improved clarity.

Because the skimmers were so effective, we were able to shut down some of the scrubbers to save labor. The lights from these scrubbers were given to the agriculture team to enhance food productivity. The crew then began an extensive process of making new planter boxes and mounting these artificial lights for supplemental lighting in shaded areas such as in the basement underneath the main fields of the agriculture.

Each light was taken out of the scrubber room, washed down to remove the accumulated salt deposits from the continual splashing of salty water, and re-mounted over the make-shift soil boxes. This increased our growing area by approximately 1000 square feet. One of the unique aspects of living in a closed system is that you are always looking for a way to use space and materials more efficiently. It was sort of like a chain reaction — once the ocean nutrient problem was solved and the valuable lights freed up, a whole new project began.

During the second year, Laser and Gaie began the process of re-constructing the marine marsh pumps. There are eleven centrifugal marsh pumps recirculating water throughout the six marsh ecosystems. They were installed before closure but continued to cause endless problems because the contractor who installed them placed the intakes too close to the sediment. What resulted was a continual intake of bottom sediments, leaves, and some small animals like crabs who unsuspectingly were pulled in by the suction of the pumps. Laser hit on the clever idea of changing the existing pipes into an air lift system. To do this he placed an air stone (just like an aquarium air stone but larger) in one end of the recirculation lines. This created a water current that ran out the other end of the pipe thus creating a flow of water without any mechanical action. All the six marsh ecosystem currents were thus driven by air! If we had thought of this years before we would have lost far fewer fish and crabs. The loss of such animals is an inherent problem in meshing pump technology (for necessary water flows) with ecological systems. It also would have saved us many hours of repair work.

General maintenance also had to be carried out on the computers, sensors, analytical equipment, and video systems. Roy and Linda helped Laser keep the many critical sensors calibrated. Frequently, Roy could be heard on the radio with Sherri Burke, the sensor specialist at Mission Control, as they calibrated the agriculture air temperature or ocean water temperature. Linda spent an hour a day working with Gary Hudman (also in Mission Control) on the air handler valve controls which proved to be a continuous problem. Taber kept the analytical and sniffer systems operational. All computers, analytical machines, and sensors had their own stock of spare parts and tools for repair, maintenance, or calibration.

Laser managed the video systems inside the Biosphere which involved a communications-conferencing system, remote cameras for viewing the wilderness and agricultural areas, entertainment (there was a set up for piping music and videos into Biosphere 2), and media interviews. The media events caused the most problem because they were complex — especially those involving satellite uplinks. Each required a camera that was set up to send an image out to satellite trucks that relayed the images to interviewers who might be in New York, London, or Tokyo. Interviewers wear remote microphones to feed their voices to the satellite truck, while providing a two-way communication back to the Biosphere through an ear piece attached to their hand held radios. Setting up for a media event would often take Laser two full days: one for setting up and one for managing the link-up and dismantling the system after its successful operation.

Other technical tasks completed during Mission One were more mundane but equally essential. Laser was on call for everything that needed fixing. Instead of calling the plumber, carpenter, TV repair man, electrician, or dish washer mechanic we would call on him. Instead of going out shopping for the latest gadget or light bulb, we would ask Laser to either fabricate it or escort us to his spare parts stores where we would find what we needed.

During the course of the year all of the biospherians became more involved with the hands-on maintenance and repair of their areas. At first, some biospherians couldn't even figure out if an electrical current had failed due to a tripped breaker! But slowly the calls for help with every little glitch started to decline, and several of us who weren't mechanically inclined actually began to fix our own problems. However, all along we knew that if we failed, 'Mr. Fix-it' could be called on as a final resort.

Aside from not having to pay huge repair bills or wait for the arrival of a specialist to come and repair equipment, there was another aspect of maintenance unique to our system. We can not afford to assume something is useless and toss it into the garbage can if it is broken because we do not have the option to go out and buy another! We had to make it work if at all possible. Electric kettles were continually fiddled with to make them work, and one of the electric blenders was epoxied back together after it had been dropped on the floor and broken into several pieces. Although battered and worn, it kept on blending sauces and soups.

✣ Alarm watch

Just like a ship, Biosphere 2 requires that someone is always on watch to make sure that all systems are operating smoothly. Also, there is always one person from Mission Control monitoring the biospherian channel on a two-way radio for automatic alarms or requests for assistance from the biospherian watch of the day. The inside watch includes a night-time check where the biospherian tours the entire system from rainforest to agriculture. Air temperatures are noted and specific checks of problem areas like the marsh and ocean water flows and waves are made. In addition, the watch double checks to make sure the kitchen stove is off. This check was introduced after the first few months because more than one of the cooks was negligent and left the daily slops for the pigs cooking on our electric stove. Our pigs were dainty eaters who would only touch their slops if they were nicely cooked. Luckily, the danger was averted by crew members who woke up in the middle of the night smelling smoke!

Biospherians always carry two-way radios. This is so we can communicate with each other no matter where we are in the Biosphere. The radios are also useful for contacting outside staff when needed. In addition, it's a safety measure in case of medical, fire, or other emergency situations. At night the radios are placed in recharger units close to where we sleep so we will hear an alarm if one should sound. For the watch person on call, the radio is the primary method used to alert others to possible problems. The radios are connected to a computer alarm system that has an 'alarm 1', 'alarm 2', 'alarm 3', 'leakage', or 'pressure' radio messages. These alarms are broadcast over the radio in a robotic voice until the problem is solved and will 'clear' once the problem is solved.

Laser was the only one who did not carry out a twenty-four-hour watch. The other seven of us took watches in rotation — we always had our watch on the same day of the week. But as the Emergency Captain, Laser was responsible for responding to all legitimate alarms that couldn't be solved by the watch person. Like most systems, ours will occasionally sound a false alarm. One time when Gaie was on watch the radio went absolutely berserk. All of a sudden there were multiple alarms coming over the radio in rapid succession. This had all the other biospherians in peals of laughter and several called over the radio, "Gaie, its your lucky day, you hit the jackpot!" Luckily, it turned out that someone in Mission Control, trying to re-set the program, triggered a series of false alarms.

A couple of weeks after closure, Mark Nelson called Laser at about three in the morning to report a recurring high-pitched noise. Thinking it might be a fire alarm, a mechanical alarm, or an indication that something was going to blow its lid, he quickly telephoned Laser for his opinion. After a few seconds as Laser listened to the noise over the phone, he concluded with a great laugh of relief that it was a cricket which had somehow gotten into a pipe or wall of the habitat!

On another occasion, Laser's voice came over the radio, "Emergency, emergency all hands to the desert near the scrubber room. Bring a stretcher and medical kit, Mark has fallen down the stairs and broken his leg." All of us went into emergency mode and one by one responded over the radio that we were on our way. Gaie grabbed the stretcher, Roy and Taber the medical box, and Sally manned the radio and telephone to keep the outside posted. The Mission Control watch called security, the on-site paramedic, and an ambulance to stand by just outside the savannah airlock. When we arrived at the scene of the accident a couple of minutes later, we found out it had been a drill! Leave it to Laser to keep us on our toes.

In the event of a real accident, four of the biospherians have been specially trained for emergency response, and all eight went through a series of drills before closure to help us prepare for potential accidents like electric shock or fires. Roy, as medical officer, was always ready to handle any accident. The outside watch is responsible for coordinating the emergency staff on the outside so that if an intervention were needed they would be ready. A paramedic is on site during the day and available at night, if required. The site also has fire-trucks to supplement the Biosphere 2 system.

In addition to medical drills, Laser also ran unannounced fire alarm drills. Fire is something that would be a major problem in Biosphere 2. Even a small fire might necessitate the crew leaving and the atmosphere being flushed out since the smoke would quickly pollute our small system. To handle this potential problem, we installed an extensive system for fire-prevention and response. It includes smoke-detectors throughout the Biosphere along with many fire alarm boxes. An automated system rings a loud alarm throughout the Biosphere in case a fire or smoke alarm goes off. The central fire alarm control panel in the command room displays where the alarm originated. Once the alarm rings, the biospherian nearest to the command room heads there to inform everyone else of where the problem originated. Those trained in fire fighting can quickly access fire-fighting equipment, which is located at half a dozen locations.

Because the glass in the spaceframe is so strong, we had to install some conventional glass panes that we could easily break in case of a major fire. The Biosphere 2 panes of glass are stronger than car windshield glass. Even a strong impact is unlikely to cause a break through the pane. During construction, a glass pane fell from the lifting crane some forty feet to the ground and it didn't even show a crack! If one of these panes breaks, there are plywood covers we can place under the glass to minimize leakage while an outside crew installs a new pane of glass.

A near emergency occurred on the night before we entered. Thousands of people wandered around the grounds eating and drinking under huge, festive white tents while laser-light beams danced on the glass and spaceframe Biosphere under a blue, starlit night. Fireworks erupted in rainbows of color. Gaie was inside the Biosphere walking down the spiral habitat staircase when she smelled smoke! She ran towards the wilderness area where it came from. Calling for help from Bill Dempster on the radio. Bill responded that he was the one who had generated the smoke by trying to incinerate algae in the scrubber room. This is one of the ways we thought of recycling the nitrogen, carbon, etc. stored in the algae tissue grown on the algae screens. Bill was using the last night to try to experiment with different ways of using this incinerator. He spent the rest of the night flushing Biosphere 2's air (which had been scheduled anyway), and the smoke was quickly gone. We were going to start the two year experiment with fresh outside air to track exactly what changed from day one, September 26, 1991, as Biosphere 2 began to diverge from Earth.

✤ *With a little help from our friends*

One of the unique aspects of our technical systems is that we can get advice and help from people on the outside. For example, support staff in Mission Control can check the computer readouts of system status and spot problems we may overlook on the inside. We can also call up the companies that made our equipment to check on the best ways to undertake repairs or to enlist their assistance if the equipment fails to perform as specified. Engineers are eager to help when they understand the call is coming from inside Biosphere 2. Companies also get highly motivated to get their equipment working up to specs when they realize that this little world is dependent upon it.

With the video system we have inside Biosphere 2, we can link up with specialists to look at problem equipment or parts. They can also talk Laser through a repair with which he is unfamiliar. A good example of how creative this repair and engineering at a distance can get is the assistance Taber received when the liquid nitrogen plant began to malfunction. This equipment extracts nitrogen from our air and liquefies it for use in the analytical laboratory. It's made by a firm outside London who had helped tailor it for Biosphere 2's needs. When we began experiencing problems with it after closure, the engineers set up a similarly modified machine in a room so they could duplicate the conditions it was subjected to — high carbon dioxide and humidity. They checked one possible cause after another. Finally, they discovered that the SF$_6$ gas we'd spiked the Biosphere

2 air with was causing the problems. They quickly worked out a solution using their machine outside London as a guinea pig. Luckily, it also worked for the device inside Biosphere 2. This was a perfect example of the power of global communications. The trickiest part of the repair by long distance was the eight-hour time difference between Arizona and England.

These challenges attracted some of the world's best and most innovative corporations, engineers, ecologists, and geo-chemists. How often are you offered the chance to build a rainforest from before scratch, from the underlying stainless steel and concrete on up? Or to recreate a miniature living ocean, moving dozens of species of coral reef from the Caribbean and Yucatan, which can scarcely tolerate even a few hours without the right water quality and movement, light and temperature? Or to engineer everything from ocean waves and tides and savannah stream flows, to making virtually airtight a structure with over three acres of glass roof, and twenty miles of joints? And all this must be accomplished in a greenhouse which recycles its air and water under a blazing desert sun in summer, and snow and wind in winter!

CHAPTER 9

Animal Tales

❧ *Domestic animals*

We spent two years living with animals — and we don't mean cats and dogs. When we first entered the Biosphere, we had no idea just how important a role animals would play in our lives during the mission. Certainly the animals had been carefully selected after years of experimenting at SBV research and development. The domestic ones were chosen for their productivity and ability to survive on the diet we had available. They ate the parts of plants that humans cannot digest such as sweet potato and peanut greens, or sorghum stalks. They turn this foliage into milk and eggs for us to eat, and their wastes are valuable for making compost, our renewable source of soil fertilizer. The wilderness animals were selected for the roles they play in the food chain. In addition to their functional roles, the animals played a vital role as companions, and even entertainers, for the biospherians.

This is particularly true of our domestic animals. We went into the experiment with three types, African pygmy goats, Ossabaw feral swine, and a chicken that is a cross between a Jungle Fowl and a Japanese Silky breed. We chose pygmy varieties because their small body size means they are more efficient food-utilizers, so for the amount of fodder available we can support greater numbers of individuals than would be possible

This goat was nicknamed Houdini for his uncanny ability to escape from his pen.

using larger breeds. Of course, these animals provide only a small amount of food in our diets — but food that is important for our morale!

Goats give nourishing and tasty milk, but everything is miniature compared to a cow. From their tiny teats, we got only forty ounces of milk per day from our four goats. Not a lot, but milk is a wonderful staple which provides the raw material for sauces, cheese, ice cream, milkshakes, and other greatly appreciated delicacies.

Our chickens were like the original wild chicken — a hardy breed that forages for insects and eats just about anything they can get. Their highly bred domestic cousins can only subsist on a diet of commercial chicken feed and have lost all their innate ability to brood and raise chicks. We did not want to use incubators in the Biosphere, so we crossed the jungle fowl with the Silkies because the Silkies are such good mothers, regularly brooding and raising chicks.

Obviously, animals need a comfortable environment in order to remain healthy. The African pygmy goats have a mountain ancestry so we provided a variety of wooden boxes to climb; the pigs were given boxes of soil to root in, water tanks in which to wallow, and any number of surfaces for scratching themselves (which seems a source of inexpressible pleasure to a pig). The chickens have overhead roosting perches, straw-lined laying and hatching boxes, and plenty of soil and mulch to scratch their way through. Roosters even have rivals they can attempt to dominate in adjoining pens. Watching them eye each other, crow, and leap up against the wire netting separating them is quite a spectacle. As a matter of both aesthetic pleasure and functional efficiency, the pens have an abundance of plants growing in window boxes and containers. They are mostly fodder plants or yams which arbor over the top of the pens, or bananas whose leaves are periodically cut for feed. The plants give the animal bay more the feel of a tropical garden than of a tightly-packed, modern animal raising factory.

Anyone who has lived near animals knows they can vary widely in personality. Our four milk does are very finicky. Having been treated like prima donnas from day one they will accept nothing but the very best. They make it clear if the snack used to lure them to the milking stall is unacceptable. Sally and Jane share the milking duties. In the morning Sally makes her way to the milking pen with the milk buckets, a saucepan with which she initially catches the milk, and a basket of goodies — usually some sweet potato greens or bean hulls — to keep the goats busily munching while she milks.

The goats have a milking order, and there is much pushing and butting at the gate of the milking stall if someone tries to go out of turn. Sheena, the largest of the does, insists on being first. Generally she will stand quietly while the milk squirts into the bucket, but if the sweet potato greens are a bit on the limp side, or she does not think there is enough food in the trough, she will give a swift kick in the direction of the milk pan. Sally and Jane grew wise to this over the months and were always ready to grab the pan quickly if they saw a kick coming.

Milky Way, the youngest doe, was terribly nervous at first and it took her about a year to settle in. She had previously been raised almost strictly on alfalfa hay and it took a while before she recognized Biosphere 2 fodder as acceptable food. Once terrified of humans, she now comes bounding up to the milking stand and no longer minds being stroked. Stardust, on the other hand, is the Mack Truck of the goat community. Nothing fazes her as she goes about her business of eating anything she can.

Vision butts and bullies the other goats and often stubbornly refuses to come and be milked. She always seems put out by the whole affair of milking and unless the snack is particularly delicious will often stand in the stall and disdainfully refuse to touch it, her nose in the air for the whole procedure. Generally the only thing that will please her is fresh banana peels, the most sought after delicacy in the goat world.

Buffalo Bill, the buck, loves to have his head scratched and goes up to the humans entering his pen and rubs up against them begging for attention. If the female goats are in heat and he is unable to get at them, he runs around butting everything in sight, making growling and spitting noises. He is very intelligent and has frequently succeeded in releasing the latches on the pens. We started to call him 'Houdini' since it seems he can get out of anything. Eventually we had to wire up the doors on his pen. All the goats, once their gates are opened, head straight for the feed bins. If no human finds them they eat until they are practically immobile — rather like a human after a traditional Thanksgiving dinner. When they are finally discovered, they are so full that they are barely able to waddle back to their pens.

We soon became intimately familiar with goat and pig health and family life. Usually the goats bear twins, but they occasionally have triplets. These births are not always without drama. Our first goat births occurred on the same day in the middle of November 1991. First, Stardust gave birth to twins before anyone arrived for the morning feeding. Then later in the morning, Vision went into labor, but the kid wasn't positioned correctly. Even with help from the biospherians it couldn't be correctly aligned. The drama went on for some hours until Dr. Barbara Paige, the vet from Tucson who has worked with us on our animal program, came out in the early afternoon with a 'demonstration kid' (one of the new-born kids from the BRDC). She tried to show Jane how to grab the kid while it was still in the uterus. She explained that sometimes the kid will just be too big, and it may not be possible to save it. But the critical thing is to save the mother's life. Jane was finally able to pull the kid out, but it was stillborn. Luckily, she was able to save the life of Vision.

In a curious twist of animal behavior, the twin kids started suckling from both Stardust, their biological mother, and Vision, their surrogate. Perhaps since both does went into labor and had milk at the same time, they weren't able to tell whose kids were whose. Since Vision seemed to be the better mother, we assigned the kids to her — to the evident satisfaction of all parties.

Sheena injured her leg — we don't know how she got the original wound, but by the time Jane noticed that she was limping, the infection was very deep. Jane and Taber regularly cleaned the wound and dressed the leg in an attempt to nurse her back to heath. At first they had to administer tranquilizers to make Sheena stay still enough to treat the sore. Sheena grew to love these shots, and would leap up on the bench that Jane used as a medical table to eagerly await the injection. Our goat had turned into a junkie! Later as the wound healed, the shots were no longer necessary. Sheena, with a slight limp, resumed her self-appointed role of 'queen of the goat pen'.

The two adult pigs, Zazu and Quincy, were captured on Ossabaw Island off the Georgia coast, an area where Gaie spent a lot of time as a little girl. The pigs there can be quite aggressive. Gaie remembers her parents telling her not to excite the wild pigs in any fashion because they may charge, and if they did she should hide behind a tree. She held them in great respect. Zazu and Quincy were anything but wild, and along with their seven piglets they provided never-ending entertainment. The piglets loved to nap inside the black plastic watering bucket, all piled up with legs and snouts sticking out everywhere.

Zazu and Quincy slept side by side in one of the wooden boxes. We called them Mr. and Mrs. Piggy. Quincy, who was the smaller of the two, was hen-pecked by Zazu. She would give him a sharp nip at feeding time if he got too pushy going after the food. One day Linda discovered that if she scratched Quincy behind his ears he would promptly go into a semi-comatose state and drop at her feet in complete pig ecstasy. A few days after closure Zazu had the entire crew on a wild hog chase when she managed to escape into the IAB having found pig-paradise in our vegetable plots, she was not going to give up easily. Frantic biospherians ran through the fields trying to cause as little damage as possible while they headed off the rampaging pig. Eventually she was returned to her pen, but not without a struggle.

Delightful though they were, the pigs began to pose some problems. We had originally worked with Vietnamese pot bel-lied pigs, very small creatures that did not eat a great deal and had such a gentle temperament, that they became popular across America as pets. Is was this popularity that eventually prevented us from taking them into the Biosphere. Animal rights activists, unable to view the animals as part of a farming system, begged us not to take them into the Biosphere and eat them. We decided to drop the pot-bellied pigs, but this meant we had to find a last minute replacement.

The Ossabaw feral swine are not as small as the pot-bellied pigs. They eat a fairly large amount of green leafy material, which was all we had available to feed them, but they were not really thriving. They would have much preferred a starchy diet, but we couldn't afford to give them any of the starches since these were a crucial part of the human diet. Also, just collecting fodder every day to keep them going was taking hours of crew time. Eventually we had to face the sad fact that we could not support the pigs in our system. They were butchered and put into the freezer to provide special delicacies for Sunday dinners and feasts for many months to come. As our agriculture area matures and provides more fodder, we may be able to consider introducing another breed of small pig into the system, but for now we have to make do with chickens and goats.

For the first year our chickens produced few eggs. This isn't too surprising since they were at the end of the line as far as food distribution was concerned. With chickens, you get out what you put in. They got some kitchen slops, though not much, since we usually ate everything on the table. They also got the cockroaches and pill bugs that we trapped, the worms Laser raised, some azolla from the rice paddies, and grain stalks from grain processing. They looked healthy, if a little skinny, but egg laying was just more than could be expected of them on that diet. In the second year, with more food production, Sally could give them scratch (the scraps left over when grain is threshed) for short periods in large doses and more trapped insects. They responded by laying a few more eggs. These eggs were highly prized items for holiday and birthday cakes and crêpes.

By the second fall we had our first three chicks hatch. When they were old enough they were moved to the goat pen where they ran happily with the goats and pecked at their slops. They were also occasionally seen hitching a ride on a goat's back. Totally tame, they could be picked up and stroked by humans without ruffling a feather. In the mornings, when Sally was milking, they would come and visit to see if any of the goat goodies were tasty. Then they would stand in the trough and peck away alongside the goat. Otherwise, they would just stand around and cackle sociably.

In order to keep the goats in milk, we have to schedule their breeding so that kids are born regularly. Once they reach a certain size the kids have to be slaughtered. The Biosphere is able to sustain only a limited number of animals. Laser was trained in the use of a stun gun because it is an extremely quick and humane way of slaughtering. The animal carcass is hung and butchered, and the joints of meat are carefully packed away in the freezer in meal sized packages. The organs are cooked up fresh for the evening meal, a treat greatly enjoyed by all the crew. Every last bit of animal is used. The pig's head and hooves are used to make posole and soups, and Sally spent several evenings in the processing room making her first blood sausages.

Galagos

Wild animals were also an integral part of our world. The stars of the animal world in our wilderness are the galagos, small primates commonly called bushbabies. One of the deciding factors to include the galagos inside Biosphere 2 was that they would provide some companionship for the humans. Even though they were raised in captivity at the Duke Primate Center and later spent time in our research greenhouses, we expected the galagos to adapt fairly quickly to being 'wild' again. In the years before closure, we enjoyed watching the galagos as they were re-introduced to the world of trees, fruit, and insects. Previously, food had only come to them via scientists in white lab coats. So what would unfold when they had access to the wilderness biomes of Biosphere 2?

Four galagos were introduced into the Biosphere. Oxide, an adult male; Topaz, an adult female; Opal, the subordinate female; and William Kim, the baby female galago born in the research greenhouses to Topaz. (The baby was named after the writer William Burroughs who urged us to include a small prosimian in the Biosphere.) There was already bad blood between Topaz and Opal before closure. Topaz had badly injured Opal shortly after her birth in the year before closure. The female brawling may have stemmed from jealousy regarding Oxide, or perhaps it was a simple assertion of hierarchy. Topaz was the alpha female, Opal the lowly beta, and Topaz didn't let Opal forget it. We hoped that the increased territory the galagos would now command might lead to some kind of peaceful co-existence.

The question of whether they would confine themselves to the rainforest, which would be rich in food and similar to the habitat of their native Africa, was answered a few weeks after closure. We were enjoying our Sunday dinner when the galagos were spotted dashing across the overhead pipes that run through the upper savannah. They stopped as if to listen in on our conversation, then continued on their way. In time this came to be known as the galago highway because we would often see several galagos heading south in the direction of the desert or north back to the rainforest.

The first inkling we got that the Topaz/Opal drama was continuing was when a night-watch spotted Opal in the technical basement. We were surprised because we anticipated that the galagos would stay in the wilderness biomes overhead. Galagos are both nocturnal and arboreal; it is thought that they instinctively stick to the trees and dislike being on the ground. In their native Africa, coming down from the trees would expose them to predators.

While performing a nightly check of the Biosphere in late October, Mark was startled when he entered the desert and heard the loud hooting of a galago. A survey of trees and spaceframe revealed Topaz in one of the tallest of the upper thorn scrub trees. Mark was answering her vocalizations, hoot for hoot, when he became aware of further movement south. To his great surprise, Mark saw Opal scrambling over the desert rocks. It became clear that Topaz was intent not only on Mark, but on her female adversary. Going to the ground was Opal's way of showing subservience to the alpha female.

Biosphere 2 is a galago's paradise. Where the trees end, there are the monkey-bars: the spaceframes. Galagos are small (an adult weighs just a couple of pounds), and their food needs are modest. But with a long tail and superb jumping ability, they're capable of prodigious leaps and can cover distances quickly. In Biosphere 2, where the trees are near the spaceframes, they leap from one to another without missing a beat. On one overcast afternoon a few months after closure, while pruning back tall grass in the upper savannah in our carbon dioxide battle, Mark and Linda were riveted by the spectacle of Kim learning the ropes in her new world. She must have been asleep in a tree above the bamboo belt that separates the rainforest from the ocean, and evidently she was awakened by the human activity. They saw her scamper along a spaceframe that runs just under the glass from east to west. Just a couple of feet from a connecting spaceframe node, she slipped and managed to grab onto the spaceframe by her powerful hind legs. While they watched entranced, Kim made a couple of efforts to swing around and reach the node with forelegs. No go. Was she in trouble? There were no vocalizations, no frenetic activity. But Linda had never heard of galagos hanging suspended upside down either.

They speculated that the galagos, normally active at night, were used to far drier and easier to grip spaceframes. During the day, and especially near the overhead glass, the spaceframes are wet and slippery, covered with condensation. While Linda and Mark lay on their backs intently watching Kim through field glasses, they began to toss rescue scenarios back and forth — tall ladders, biospherians with safety lines climbing the spaceframes to reach her some thirty-five feet above the ground. There was a lot of vegetation below the spaceframe so even if she slipped she might be able to grab an acacia branch (ouch! the thorns!) to break her fall. Finally some twenty minutes later, a few quick swings around the spaceframe brought her nearer to the node, and with an assist from her tail, she was back upright. A split second later she had decamped onto a nearby Leuceana tree — as if she only trusted the natural branch now that she'd seen how slippery this steel tree had proved to be. The sun was setting. Mark and Linda let out a collective sigh of relief — no need for the rescue team tonight.

In mid-October after Roy had spotted Opal in the tunnel that leads to the south lung on his night watch, Linda came down to lure her with a banana. Linda had nursed Opal after she'd had her row with Topaz, and was closer to her than anyone. Sure enough, step after cautious step, Opal finally came within reach of the banana. Linda was concerned that perhaps Opal was lost in the technical basement of Biosphere 2 and couldn't find her way back to the biomes overhead. Linda placed her back into the galago cage in the lowland rainforest, where she could be fed for a while. Kate Izzard, the galago expert at the Duke Primate Center, counseled that Biosphere 2 should be large enough for two females to stake out different areas. So early in November, the cage door was left open for Opal to make a second attempt at life in the wild. While Linda watched, Topaz came in for her food set out on the cage top, and stood guard from a nearby tree glowering at Opal who crouched fearfully inside. When dawn broke the next day, Linda took Opal out of the cage. Opal would have an hour or so to find herself a comfortable sleeping spot before the morning light signaled her it was time to call it a night.

After that, we began to be accustomed to seeing Opal in the basement. For a while, she slept near the ceiling of the room that leads up to the room with the algae-scrubbers. Linda put out her chow in the technical basement to ensure that she had access to food to supplement what she was able to forage upstairs.

One afternoon, Roy was working on calibrating sensors, and had discovered Opal curled up asleep inside one of the boxes that relay signals from our video cameras! When Linda reached Roy, Opal had partly awakened, and was regarding them through sleepy, half-closed eyes. Since the box contains no live current, she was left to enjoy the rest of her sleep. Later Gaie saw her by the ocean, sleeping in a plastic bin on the diving platform!

Early in March 1992, we celebrated the birth of our first Biosphere 2 galago baby. Oxide was very protective of her, and it appeared she was, by the time we saw her, at least a few weeks old. But since the gestation period for a galago is about four months, she was very likely conceived inside Biosphere 2. Although this did little to appease the interest of those journalists who persisted in asking us whether a human baby was likely during the two years, it marked a certain milestone in terms of the acceptability of our artificial world by our evolutionary relatives.

But a tragic accident occurred in late March just a few days before we celebrated our first six months inside Biosphere 2. William Kim had evidently also taken to exploring in the technical basement, and died of an electric shock when she touched a power transformer box which had a small opening virtually invisible from the floor below. Deeply saddened by the loss, Laser and Linda spent time the next few days meticulously checking and safeguarding against any other dangers that the tiny fingers of our galagos might somehow encounter because it was clear they possessed the curiosity to roam the parts of Biosphere 2 most remote from their rainforest habitat.

In the meantime, Opal's drama was far from over. Despite Opal's retreat to the dungeon, the technical basement, Linda found her in the summer with an ugly-looking scalp injury, probably the result of a scrap with Topaz. Fortunately, Roy's X-rays showed no bone damage. We kept her in a large cage in the medical room for some weeks while we administered antibiotics and cleaned her wound regularly. Then we dismantled the galago cage from the rainforest and moved it into a shady corner of the orchard (on the agriculture side of the Biosphere) so that Opal would not have to endure harassment from Topaz. But when we made her new room ready, Opal found a tiny gap in the overhead mesh and announced her escape by hooting from the giant banana trees in the orchard. By the time the hole was located and patched, Oxide had joined Opal by slipping through loose screening that divides the savannah from the orchard.

They refused to be lured back in, so then came the fun of figuring out how to trap them. We finally came up with a long string attached to the door of the galago cage. Baited with ripe bananas and monkey chow, biospherians took hourly turns discreetly seated in a wooden chair some twenty to thirty feet away. The trick was a quick release of the door while the galago was in snacking. It worked. Oxide was trapped and returned to the rainforest. Opal was reintroduced to her tree-shaded cage.

But no sooner did we think order had been reinstated than the familiar sounds of night hooting from orchard trees and IAB balcony told us that the galagos from the wilderness were determined to have the run of the human areas as well. We had been afraid of damage that galagos might cause if they had access to the IAB, but careful checks revealed they limited themselves to occasional nibbles at the ends of nearly ripe bananas. They preferred heights, and showed no interest in clomping through the agricultural plots and rice paddies. It seemed we could live with the arrangement, and in fact we have been delighted by the occasional galago conversation late at night when we go out to stroll and lean over the IAB balcony. The only place that our irrepressible galagos haven't successfully made part of their domain is the human habitat. We often joke, though, that one of them may turn up for a seat at the dinner table.

Thereafter, Opal seemed content — safe from her female nemesis. She is slated to rejoin our growing galago colony in the research greenhouses after the two years. Like us, she will have her own unforgettable memories of being a participant in the first two-year closure of Biosphere 2.

✌ Stowaways

There are several types of animals and insects that were not deliberately introduced into the Biosphere, stowaways who refused to leave. One such was a large curved-bill rock thrasher who hung out in the savannah, thornscrub, and desert. At first we tried to snare it and remove it, but it was wise to the bird nets we'd used to catch most of the English sparrows before closure for release outside. In addition, the thrasher had a lovely, plaintive song. So we eventually decided that it was not doing much harm, and since the birds that we had deliberately introduced had not done well, we decided to leave it. It was nice to be able to hear bird songs in our wilderness, even if it was not exactly the bird song that we had originally planned on.

The same did not apply to the IAB where some sparrows were competing directly with our stomachs. There, a large flock of sparrows had also moved in during construction and lived happily on what they could steal from our grain crops. We tried everything we could to trap them and get them out. We caught the majority in nets before closure but three very shrewd birds evaded us. One by one they died off until one lonely sparrow was left by the second year. We gave up trying to catch it. We often wondered how it felt, being the only sparrow left in its world.

Mice live in our basement, and probably all over the wilderness biomes though they are not so visible there. In the IAB basement we trapped large numbers of them because they could really become pests eating our precious grain supplies. We were careful never to leave any grain where they could reach it. We also trapped large numbers in the habitat area. They seemed to have free run of the habitat via the electric conduits. We tried various different baits. The trick was to get something that they could not remove from the trap without triggering it. Sticky banana mixtures seem to work well, but then the problem was keeping the cockroaches away from the bait that was intended for the mice. Yes, even in Biosphere 2, man is engaged in his age-old battle against the mouse and the cockroach.

It seems from our experience that interaction with other members of the animal kingdom is an important part of life for human beings. Toward the end of the second year we began to evaluate whether or not we should keep galagos in the system since it would still be some time before the wilderness would be mature enough to support them. Everyone agreed that it is worth the trouble of taking care of them simply for the pleasure of watching their antics. The same applies to the domestic animals. We could have devised a diet completely free of animal products and found some other way of composting our inedible plant material, but the domestic animals enrich our lives as well as our diet.

CHAPTER 10

The Three-Acre Test Tube

✣ A unique laboratory

Geophysicist Keith Runcorn is a Fellow of the Royal Society of London who has made significant contributions in the area of plate tectonics, helping to confirm that continental drift is a reality. At the Environmental Symposium held at SBV the day before closure, Runcorn pointed out that while physical scientists have become used to the power of advanced technological tools, scientists who study the environment have very few such tools. Moreover, ecological studies often suffer in the competition for scientific funding, despite the obvious importance of the global environment. This made Biosphere 2, because of its scale, complexity, and data production, a unique laboratory, the first of its magnitude and kind in the arena of life studies. We could not know in advance what we would learn by working inside it, but as Runcorn pointed out, new scientific tools let you see what has never been seen before.

Our two-year mission would, among other things, be a shakedown voyage where we would learn to operate an entirely new kind of scientific laboratory. Since we were at once carrying on research activities and learning to use the facility, the border between research and operations would be crossed back and forth many times in the first two years. For example, data about water quality and air composition is collected for health and

Measuring the growth of trees and grasses is an important part of the ecological research conducted inside the facility.

safety reasons, but it also provides new information on how these cycles work. In the agriculture system, it could be said that in a sense there is almost no division at all between operations of the facility and the research about sustainable agriculture and chemical-free integrated pest management. To a great degree, the operations are the research, and the research informs the operations. Our primary intention, beyond checking the biospheric hypothesis itself, was to observe the unexpected, and with it, seize the chance to study something radically new.

There were times when we argued at the breakfast table about how much time within the parameters set by mission rules we should allocate to research and how much to needed operations. At one point in the first month, Laser erupted, "Is there going to be a split here: scientists versus technicians?" There were several biospherians who at first were quite unhappy that we didn't have more time for direct research. We had all agreed to participate in the first mission knowing full well that all of our time might be used dealing with operational necessities. After all, there are bugs in any new system. But the exact division of research and operations was a critical point and one that would be repeatedly discussed. Without Biosphere 2, there was no study of biospherics, hence our number one objective was to keep all systems operating. But without research, the purpose of the Biosphere would be diminished.

As we became increasingly efficient at operating Biosphere 2, more time became available for pursuing specific research. By the end of the first year, there were over sixty scientific research projects underway. Although there were only eight of us inside, there was the SBV research staff outside to whom we reported, and we had hundreds of scientific and technical collaborators,

The Herculean task of cutting back the dormant grass in the savannah is capped by a moment of light-hearted comraderie.

PETER MENZEL

(above) Linda measures inter-nodal lengths on a desert plant.
(right) In the rainforest, she reseeds an area of the cloud forest
mountain.

MARK NELSON

D.P. SNYDER

Gaie and Laser
document the
health of the
coral reef with
an underwater
video camera.

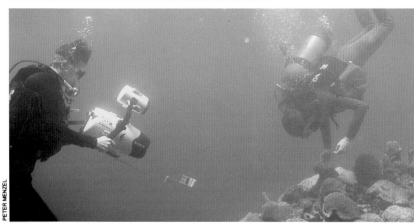

PETER MENZEL

(right) The structure is so vast that communications equipment needs to be close-at-hand.
(below) Mark collects samples for leaf-litter studies in the marsh.

(below) Laser machines a new part in his workshop.

*(top) Biosphere 2.
(left to right) Lower
thornscrub, savannah,
rainforest, ocean, desert.*

TOM LAMB

TOM LAMB

GONZALO ARCILA

GILL C. KENNY

(above) Taber and Jane formed the basis of the Biospherian Band. (left) Laser and Gaie share a quiet moment in the tower library.

(right) Laser has turned his apartment into a video studio.

(above) Dr. Oleg Gazenko and John Allen exchange a biospherian handshake with Mark Nelson.

(above) Linda, Jane, and Taber share a hearty laugh. (right) A beach-blanket birthday celebration.

PETER MENZEL

GILL C. KENNY

(top) The crew of Mission One contemplates re-rentry to Earth's biosphere.
(bottom) As a research replica of the mighty planet, Biosphere 2 has its work cut out for it: to provide information for the intelligent management of the environment on Earth and to lay the foundation for living habitats in space.

ranging from consultants, to friends, to those who simply had an interesting suggestion to make. Scientists on the outside sent and received information along computer cables, across telephone lines, and via video image. We used state-of-the-art computers, sensors, and multi-media facilities to create the possibility for SBV staff and outside scientists to join us on an ongoing basis in areas of special interest such as gas cycles, soils, and physiology.

To provide information for both current and future researchers who may want to track small changes in environmental parameters, there were nearly 2,000 sensors installed in Biosphere 2. These electronic sensors allowed us to measure variables like carbon dioxide, pH, temperature, and humidity while analytical equipment makes other detailed water and air analyses possible. Each sensor is designed to analyze one variable and its data is then connected to a series of command and relay computer stations that make up the complex nerve system of Biosphere 2. The data they generate is archived at Mission Control in duplicate form for safety. Since these sensors generate data either continuously or on a fifteen-second or three-minute averaging, that adds up to a lot of data — about 10 megabytes per day!

Some have called Biosphere 2 a 'cyclotron for ecology', and in some ways it is an appropriate analogy. Just as a cyclotron speeds up the motion of atoms to allow physicists to discover new subatomic phenomena, Biosphere 2 speeds up certain ecological cycle, and offers new ways to study the fundamental processes of life on Earth. Carbon dioxide enters and leaves our atmosphere thousands of times more quickly than in Earth's vast atmosphere. Water moves from the ocean, to atmosphere, to drinking water, to raindrops in a period of weeks. We lived by the rhythm of these accelerated cycles, and one calendar year of study in Biosphere 2 allowed us to collect better controlled and far more data that we could have in the same year in the Earth's environment. Of course, one interesting difference is that the cyclotron pursues the investigation of the smallest physical particles and Biosphere 2 that of the largest life systems.

❧ *Carbon budget*

Among the most potentially important of the research projects carried out was modeling the carbon cycle. Where was all the carbon when we closed, and how will it change over time? This may give us insights into subtle mechanisms at work in Earth's carbon cycle — now a hot topic of both scientific study and public concern because of the threat of global warming.

Ten percent of the approximately five billion metric tons of carbon added to the Earth's atmosphere each year is unaccounted for — we just don't know where it's going. The oceanographers claim that the carbon is being stored in terrestrial biota, but the forest ecologists say that this is impossible. They argue that because of the increased deforestation world-wide, carbon is being released by forests faster than it is being taken up in biomass. They point their fingers back at the oceanographers: the missing carbon has got to be in the ocean waters. And there are some scientists who think that since all the numbers we have are so imprecise, there may not be any missing carbon at all!

Carbon moves through both living and non-living portions of the biosphere (air, water, soils, rocks, and living organisms), but never before have the actual flows, rates, and amounts been measured within the framework of a finite, closed, ecological system. We can accurately determine these things in this complex world of Biosphere 2, which includes many of the most important components of the global biosphere. Even the rates of movement are not static. They depend on temperatures, light levels, rainfall, humidity, carbon content and molecular structure, nutrient availability, and so on. Carbon models used today by scientists and government officials lack supporting data and are over-simplified, steady-state diagrams. Therefore, the real carbon situation in the world today is much in dispute.

Knowing the sources and sinks of carbon is critical if, for example, we want to make any predictions about the effects of global warming (caused by increased levels of carbon dioxide and methane in the atmosphere). Over the past two hundred years or so, levels of carbon dioxide in Earth's atmosphere have risen from about 280 parts per million to 360 parts per million. Most of the increase in carbon dioxide is caused by the burning of fossil fuels, but a certain amount is also attributed to the worldwide destruction of forests and other ecosystems. Still, we know little about the sources of carbon dioxide and even less about where carbon dioxide is deposited or transferred. Without dynamic and accurate models of how carbon cycles work, even the bare facts of global warming (not to mention the implications for human technology and ecological health) will continue to be unknown. So, appropriately, measuring the effects of the rising carbon dioxide on vegetation and corals was the first of the great opportunities we had to work with in our two-year experiment.

We started off by tackling the question of how much

carbon resided in plant tissue. Before closure, surveyors mapped the entire Biosphere, identifying the exact location of almost 10,000 plants. During the two years, we re-measured subsets of these plants to get an idea of how they were growing. Using standard techniques and a few simple measurements (the diameter at the base of the plants, the diameter at breast height if tall enough, overall height, and spread of the leaf canopy), we estimated the amount of biomass and carbon each plant contained. From around eleven metric tons of biomass when we closed, we had increased to around twenty metric tons judging from the partial resurvey we conducted in July 1993.

To determine how plant growth in Biosphere 2 differs from 'normal' because of our lowered light, higher carbon dioxide, and lack of strong winds, we will re-measure every plant during the transition and do a limited amount of destructive sampling (that is, some plants will be cut and dried to determine carbon content and density). Since many of the first canopy trees that have fulfilled their function of providing sunlight protection during the early growth of the rainforest community, they will be cut down during transition and used for this study. These measurements give us valuable data about how our ecosystems mature, how elevated carbon dioxide affects their growth patterns, and which types of plants dominate.

Another key to the model of carbon flows is the soil reservoir which contains over ninety-five percent of the organic carbon in Biosphere 2. Despite its importance in all ecological processes, many of the fundamental processes of soil formation and evolution are still little understood. This is not so surprising given that every gram of soil contains millions of bacteria and fungi, each contributing a diversity of functions. But therein lies the power of soils — they are important agents in ensuring that nutrients remain available to other forms of life.

To sustain the wide variety of special habitats inside our world, our soil geologists created more than thirty different soils or marine sediments on the model of Earth's soils. Some are twelve to sixteen feet deep to support the massive growth of trees over the coming decades. We have a great opportunity to observe the changes that will occur in the soils over time. To do this, a complete set of soil samples was taken before closure, and then the soils were resurveyed at periodic intervals. Chemical analyses, in particular analyses of carbon, nitrogen, and phosphorus, will show us how these soils are changing. And from this, we can begin to understand how the different vegetation groups they support and our seasonal patterns influence them.

When news of Biosphere 2's elevated carbon dioxide reached researchers in the United Kingdom, William Chaloner, the head of England's International Geosphere/Biosphere Program, and David Beerling of Sussex University had reason to be especially interested. They are paleobotanists and have developed techniques for studying how plants in Earth's earlier history have responded to carbon dioxide levels. They study ancient plant leaves preserved in fossils and count the density of stomates (openings in the leaf for air and water). These can then be correlated with geologists' estimates of the concentrations of carbon dioxide in Earth's atmosphere millions of years ago. Biosphere 2, they saw, would give them a chance to check their work by comparing the stomatal density of a wide variety of living plants exposed to the type of extreme fluctuations in carbon dioxide that only occurred on Earth over a span of hundreds of millions of years.

To support these studies, we collected plant samples, exported them out of Biosphere 2, and shipped them by express mail to laboratories in England. These studies may provide information about the type of plant responses we may expect if global carbon dioxide continues to rise.

This is only the beginning, because rocks, water, and animals also contribute to the flows and fluxes of carbon. Therefore we have to monitor the various aspects of the environment until all possible sources and sinks of carbon are meticulously tracked. Because we have a closed environment and a great deal of life activity, we can track carbon dynamics through the tendency of plants to use some naturally occurring isotopes of carbon preferentially. From these data and the continuous monitoring of our carbon-containing gases, we will create a dynamic computer model of Biosphere 2's carbon cycle.

❧ *Where was the oxygen going?*

News of the oxygen decline attracted Dr. Wallace Broecker and one of his graduate students, Jeff Severinghaus, at the Lamont-Dougherty Laboratory of Columbia University. Dr. Broecker agreed with our initial assessment that the oxygen was probably being lost in the soils, although the ocean and even the concrete basement and rock work were also suspect. But which soils and by what mechanisms? He targeted one soil type, Wilson's Pond soil, which we had introduced into the agriculture and in a few places in the wilderness. The soil was taken from a pond just north of Biosphere 2 that was frequented by cattle and therefore was rich in organic material. Broecker theorized that it was also probably largely anaerobic (without free oxygen), since part of the year it was flooded by monsoonal rains, and in its new environment in Biosphere 2 would be an ideal sink for oxygen. There were many other possible mechanisms that could be at work as well: the oxidation of soils containing iron or sulfur compounds (when a soil becomes oxidized it absorbs oxygen); the denitrification of nitrogen; and the formation of caliche (calcium carbonate minerals) in specific soils. With some 30,000 tons of soils inside Biosphere 2 comprised of about thirty different soil types, it would definitely take some detective work to track down the culprits.

Concrete is, in essence, calcium hydroxide — the same chemical which forms a key part of our chemical recycling system for carbon dioxide. Calcium hydroxide plus CO_2 makes calcium carbonate (a powdery form of limestone). It is known that CO_2 reacts mildly with concrete, penetrating it shallowly. But in Biosphere 2 with its elevated levels of CO_2, its vast surface area of simulated rock cliffs and mountains, its cement floors and pillars, could it be a significant player in the oxygen mystery? And if it is, what ramifications does concrete have for oxygen levels in our Earth's atmosphere? To address this question, an additional twelve one-foot-long concrete cores were taken from cement walls or rocks throughout the Biosphere to analyze of the composition of the concrete. And indeed, when these analyses came back, our mystery was solved. In the higher CO_2 environment of Biosphere 2, about ten times as much CO_2 was being absorbed by the concrete than would occur in Earth's atmosphere. So the oxygen was being tied up by first oxidizing some of the organic matter in our soils, then the released CO_2 was being absorbed by the concrete. From the carbon isotope analyses, the signature of the CO_2 in the concrete was a perfect match for what we were looking for.

There is an interesting lesson to be learned from this: even in a small man-made biosphere, the complexity of interactions combined with our ignorance of basic information about how ecosystems and biospheres function makes surprises and discoveries inevitable. In a sense, noticing the unpredicted oxygen depletion and finding its surprising cause were the kinds of discoveries the Biosphere was built for.

As a spin-off of the oxygen question, Wally Broecker introduced us to two scientists, Dr. Martin Wahlen, Professor of Geochemistry at the Scripps Institute of Oceanography, and Dr. Mike Bender, Professor of Geochemistry at the University of Rhode Island, who are interested in monitoring both oxygen and carbon isotopes. An isotope of an atom is a form of a chemical element that differs from the regular element in its atomic mass. For example, the most common form of oxygen has eight protons and eight neutrons, giving it an atomic weight of sixteen. One of its isotopes, O_{18}, has eight protons, but ten neutrons, making it slightly heavier. Isotopes of carbon and oxygen are present in Earth's atmosphere, water, and biomass at specific and fairly constant ratios. Will they be found with the same ratios in Biosphere 2? How do they behave? Gaie and Mark began collecting fresh growth of plants of different types from around Biosphere 2 every month for analysis of carbon and oxygen ratios. In addition, periodic air samples were sent out for these studies. Our preliminary findings suggest that the amount of O_{18} in Biosphere 2 air is higher, due to the action of photosynthesis and plant and animal respiration. In addition, patterns of how the other plants are favoring one isotope over another are starting to emerge. Through the use of isotope analysis, we will be able not only to identify the mechanisms involved, but also date when the reactions occurred because Biosphere 2's carbon and oxygen isotope ratios are different from Earth's and are steadily changing over time.

One article published about Biosphere 2 stated that the loss of oxygen was due to the fact that the humans were consuming it all — an incorrect idea, since the oxygen decline is on the order of 1000 pounds of oxygen every month. Eight human beings couldn't possibly consume this much oxygen. Indeed, all the oxygen humans consume is released as an equal amount of carbon dioxide. This carbon dioxide is then disassociated by photosynthesis of plants and algae. Atmospheric oxygen and carbon dioxide are inextricably connected.

Some of us had never studied of chemistry, and for the first time in our lives the world of atoms and molecules had become an essential part of our thinking. Chemistry was no longer just an arcane discipline studied in textbooks or laboratories. It was integrally linked to our well being. Finally, in August 1993, the brilliant detective work of Jeff Severinghaus, Wally Brocker, Bill Dempster, Martin Wahlen, and Mike Bender led Jeff to the discovery that most of the oxygen was being trapped as calcium carbonate in the concrete that protected the stainless steel liner around the floor and basement walls of Biosphere 2. This gives rise to thoughts about the effect of increased carbon dioxide on Earth interacting with our expensive concrete infrastructures.

✌ *Ocean research*

In studying the ocean, a team of marine researchers including biogeochemists, microbiologists, ecologists, and coral experts work with Gaie in tracking the health of the largest man-made coral reef system ever created. Coral reefs have suffered greatly and are under enormous pressure from human activities. Because the mechanisms responsible for their ill health are far harder to determine, the detailed environmental studies of water quality, light, and temperature which are possible in Biosphere 2's coral reef can be especially useful in determining what factors are critical to coral vitality.

The exuberant Dr. Phil Dustan, at the College of Charleston, an authority on coral reefs, joined the marine team in 1992. Phil has spent years studying the Florida reefs, noting their deterioration over the years primarily due to human impact. He was excited by the opportunity to monitor the health of a carefully measured system of corals through the two-year experiment and beyond. Corals are the key indicator species of the reef's health, and their preservation in Biosphere 2 is one of our major challenges, just as in Earth's oceans.

Phil has developed a marvelously effective system of monitoring the health of corals using a video program. Humans can be quite subjective in their assessment of how corals look, but using his video system, we were able to send Phil images of the ocean from before and after closure to track all the coral colonies. Using a color stick and ruler against each image, it is possible to quantify the health and vitality of the corals by monitoring minute changes in the color of their tissue as well as observations about their overall health. For Gaie or Laser working alone to have done such rigorous surveys and recording of information would have been impossible given our work load, but now Phil and his graduate students make these observations from the video tapes.

Phil Dustan's participation was particularly timely, because it was in March 1992 when Gaie noticed the first signs of possible coral disease: a white band encroaching the outer rim of some of the brain corals. Corals had had mixed success since their introduction almost two years earlier. Some had thrived, some had died, some struggled. Bleaching occurs when the algae that live inside the tissue of a coral die. (It is the algae that give coral its colors.) It is possible for algae to reinhabit the tissue at a later stage and return the color to the corals. There seemed to be no particular pattern to why some individuals were in excellent condition and others were not. This is exactly the question that we wanted to track in more detail.

The peculiar white band that appeared overnight looked nasty. Phil confirmed that this was a bacterial disease called, appropriately, 'white band disease' that he commonly saw in Florida but had not seen before in any of the videos that we had been sending him. Somehow, the bacteria must have been lying dormant ever since we had introduced the corals. Now the bacteria were actively at work.

We had sailed through the first six months without any problems with the corals. Previously, we had anticipated that winter would be the most difficult time for the reef because of low light and high carbon dioxide levels, but it was the springtime that sparked a new shift. The phenomenon, contrary to our expectations, coincided with increased light levels and raised water temperatures (from seventy-six to eighty degrees).

Later that spring, the conditions of the ocean began to shift again — this time to an overabundance of algae: there was a small chlorella algae (we called it the 'grey slime') that covered everything, bottle green algae that grew around the hard corals, red spiny algae that grew among the soft corals, and a calcareous red algae that grew along the lagoon floor and reef rock. These four algae species in particular had become terrible pests and continued as significant weed problems throughout the two-year experiment. These algae are quite different from the tiny algae that actually cohabit the coral tissue. These macro-algae, if left alone, would inevitably cover the coral tissue and prevent the corals' algae from receiving enough sunlight and the coral polyps from ingesting the plankton it feeds upon. Like Linda and Mark in the terrestrial ecosystems, Gaie had to periodically weed the ocean system in her role as keystone predator.

We suspected that the shifting light levels from season to season could be a fundamental cause of their stress. Corals are not used to adapting to such changes, and it must be particularly difficult once winter gives way to the bright sunlight of spring. To further address this question, Phil sent us a light sensor to compare with the one we had mounted permanently underwater, he also conducted some comparison studies with the reefs of the Yucatan of Mexico, where most of the corals had been collected, and along the reefs of Belize. Indeed he found that light levels in late spring may be too bright for many of the corals that had done well in the winter. We then began to consider if our reef should have its own distinctive pattern: a migrating coral colony that moved up the reef in the winter and down the reef in the spring. To date we have not yet followed up on such a strategy; instead we focused on the necessity to increase the quality of the water by fabricating and operating the new nutrient removal devices, the skimmers, to replace our original algae scrubbers. This seemed to take care of most of the problems as we continued to track the corals over the next year. In March 1993, there was another slight sign of stress with the arrival of spring light levels, although nothing like what had been experienced the year before.

Gaie and her colleagues started a coral reef research center in Belize to study a natural reef in conjunction with our artificial reef. This center gives us the opportunity to compare vectors like the health of coral species, the chemistry of the water, and the diversity of the microbial community which forms the base of the system. One of the consultants, Dr. Donald Spoon, a professor at Georgetown University, conducted a preliminary study in January 1992 where he found eighty-two species of microbiota in the unpolluted lagoon of Calabash Island in Belize as compared to seventy-five species in the lagoon of Biosphere 2. What this tentatively indicates is that the microbial population of our little lagoon remains remarkably diverse even though it is far smaller than the unspoiled wild lagoon in Belize. For over two years, the population of microbiota has not diminished inside the artificial system even though it was subjected to mechanical pump action.

Phil Dustan and Dr. Judy Lang, a coral biologist at the Texas Memorial Museum, are particularly interested in tracking individual corals over time. Phil started a baseline study of the corals in Belize to set the initial parameters. Their past years of research in the Bahamas showed that individual coral colonies of the same species have different responses to the environment and that the ability of corals to thrive in Biosphere 2 or in the Bahamas seemed to be specific to the individual and not to the species. This was not obvious, in fact, even counter-intuitive, when we started two years ago. The initial consultant to the marine systems had advised us that we would lose at least thirty percent of our species; others had suggested even greater losses.

But as of the summer of 1993, Phil was still tracking over 500 colonies of corals in Biosphere 2's ocean with only one species — represented by a single individual — lost during the two years. Instead of vast species extinctions, we are finding that individual variation is critical. Some of the colonies continue to flourish while others show a slow reduction of tissue. It was this individual response that we wanted to understand in more detail and which will form a basis for future research.

Dr. Bob Howarth and Roxanne Marino are a delightful team from Cornell University working in marine ecology and chemistry. They manage to bridge the worlds of the pro-active environmentalist groups and the academic university systems. Bob is well-known for speaking on behalf of Greenpeace environmental crusades with indisputable scientific fact at his command. He is one of the world experts in the impacts of oil pollution on marine systems and editor of the journal, *Biogeochemistry*, his other area of expertise. Roxanne helps Bob run the laboratory at Cornell. Together they are organizing a program for tracking the marine chemistry dynamics and biogeochemical cycling of nitrogen and phosphorus in Biosphere 2, which will be developed to include similar observations at the pristine Turneffe Atoll reef community of Belize.

Of course, all marine research projects cannot really be separated from one another. Each specialty is an aspect of a total system, not an independent reality in itself. To reverse harmful trends, we need to learn more about the precise mechanisms which lead to vitality or disease in corals, and then develop appropriate means of intervening to restore the ocean's health.

❧ *Leaf litter and decomposition*

In the marsh, Matt Finn, a doctoral student at Georgetown University and research ecologist for SBV, heads the major research program. His thesis compares the ecology of the natural Florida Everglades marsh system with its replicate marsh in Biosphere 2. The Biosphere 2 marsh was designed to represent an area that spans over 100 miles from fresh water inland marshlands to the offshore red mangroves, and indeed our relatively tiny 4,200-square-foot marsh is extremely diverse in its community structure. It was packed with species which have not only been maintained but have grown several times their height since their introduction. This system is as complex and diverse as what you would find in the real wilderness.

One of the key studies initiated in the marsh, and then expanded to include all the terrestrial wilderness biomes, was the study of leaf litter and decomposition. Litter, to an ecologist, is quite different from the unwanted junk clogging up your gutters. It's the fine rain of leaves, twigs, flowers, and fruit that plant communities drop on their soil during the year. The accumulation of litter is the path to nourishing the soil, replacing what is taken out. About 130 square, mesh litter 'baskets' sit on short plastic legs in rainforest, savannah, thornscrub, desert, and marsh patiently collecting the material that rains down from above.

At the end of each month, Mark, our resident litterbug, moves on hands and knees with a small whisk broom and dustpan collecting the material from each litter basket. In the marsh, the litter baskets sit on taller legs, so they are above the water level, and the litter must be collected while wading through the water. In this ecosystem, litter falls into the water and floats until it slowly decomposes and sinks to the bottom sediment.

These bags of litter, one for each of the 130 sites, are sent out through the airlock with the monthly exports to be dried and weighed in the outside labs and then chemically analyzed. From these studies, we get a microscopic view of the topsoil — the interface between the life below and above. They also show us how each biome differs in its nutrient cycling. The numbers reflect either the large tropical leaves and branch prunings of the rainforest or the scattering of litter islands found under desert shrubs.

We also set up a decomposition study to complement the litter study. In each ecosystem, mesh bags were made with identical weights of leaves from the dominant plants of that system. These were sealed and placed on the top soil or sediment on September 10, 1992. There were twelve such bags made for each area so that each month for one year, one of the bags is collected and analyzed. This gives us the ability to determine how fast the material disappears and is incorporated into the soil. In the marsh these bags were placed in the system and tied down with fishing-line so that they wouldn't float away.

Matt is doing the exact parallel study in the Everglades with Dr. Pat Kanjas of the University of Maryland. Over the coming years, he will study how our young and growing marsh conducts its nutrient circulation alongside its parent marsh, which is subject to hurricanes, strong tides, human pollution, and theft of water. The Everglades may be the first national park facing the danger of dying as an ecosystem. It is threatened because farm irrigation diverts the fresh water that normally flows through and replenishes the marshes. Its other enemies are the chemical residues that drain through it from fertilizers, and herbicides and insecticides now affecting the delicate aquatic food webs.

❧ Wilderness studies

One of the most interesting things about Biosphere 2 is how it maintains relatively small populations of plants and animals. This is invaluable data because it parallels concerns about biodiversity in Earth's many ecosystems where human impacts, such as road building and forest clearing, have created smaller habitat islands where smaller populations of organisms struggle to survive. Among the problems this creates are genetic 'bottlenecks', when smaller populations descend from a handful of ancestors. Some baseline studies of genetic diversity in insect, fish, and plant species were conducted before closure and will be repeated at re-entry to determine whether such inbreeding is causing problems.

To further these studies, we collected a sampling of insects by setting out jars with rotting fruit as bait. The insects are freeze-dried in the medical laboratory inside Biosphere 2 for later examination by experts. A number of individual plants of several species in each biome had leaves collected for genetic analysis before closure and repeat examinations will be done during transition. Two species of a fresh water fish, *gambusia*, that were stocked before closure to observe if one species outcompeted the other, were trapped at the one-year anniversary and frozen for future study.

To track insect changes since closure, we used several

techniques to capture and export individuals for identification by Dr. Scott Miller of the Bishop Museum in Hawaii, who has been working with the project from its first design stages. Every month, pitfall traps (deep plastic containers dug in so that their lip is at soil level) are opened for twenty-four hours. Insects that fall in are collected, preserved in alcohol, and sent out for identification. For night-flying insects, we hung black lights in front of a piece of white cloth and collected the insects that were attracted. To monitor insects which reside in leaf litter, we used a special device called a Burlesi funnel where a low-wattage light creates heat to drive the insects down through the mesh of a funnel and to the bottom of the bucket where they are preserved in alcohol.

To check on how our plants respond to their somewhat unusual environment, we also did a monthly check on their flowering, seeding, and overall growth. Linda collected detailed data in the rainforest, savannah, thornscrub, and desert, and Mark sampled the marsh, beach, and the salt-loving plants that grow in the halophyte garden on top of the wave generator. During these repeated observations, we recorded notes on insects and animals, as well as making other observations. One long-term research activity is correlating the phenology of the plants with trace gas dynamics in our atmosphere, as plants release a variety of minute gases during their growth and flowering. To study what is released into the atmosphere, we collect monthly samples of pollen which are sent to Dr. Mary K. O'Rourke at the Respiratory Science Center of the University of Arizona for analysis.

It was a common sight to see SBV's desert consultant, Dr. Tony Burgess of Tucson's Desert Laboratory, wandering around the outside of the Biosphere as he makes his monthly observations, which are then compared to the ones we make inside. Tony is a self-declared desert rat, a colorful character with a distinctive bushy red beard. He is extremely knowledgeable about the ecology of the desert and thornscrub ecosystems and is carefully tracking the evolving community structure of these biomes.

❧ Sensory studies

Our two years inside gave us the opportunity for some unusual medical research. Working with Dr. Gary Beauchamp, Director of the Monell Institute, a leading institute in human sensory research, Roy initiated experiments to observe changes in our sense of smell. By using 'smell' bottles which contain a variety of concentrations of two distinctive smells, roses and spoiled milk, they are trying to ascertain if we had a reduced or enhanced smell capacity. To do this, Roy gave each crew member a whiff of a bottle with a smell odor and one without any fragrance to test our sensitivity to the odor.

To monitor our overall level of stress, Roy collected a series of daily urine samples from each of us which was examined for the hormones associated with stress. Urine was also collected early in our adaptation to the calorie-restricted diet and sent out for analysis to ensure that we were receiving adequate amounts of protein.

Another intriguing study is being conducted by Dr. John L. Laseter, Director of Accu-Chem Laboratories in Texas. Laseter is looking at what environmental compounds are present in our bloodstreams. During the two-year closure, periodic blood samples were sent out for analysis. One early finding of interest showed that levels of pesticides and other toxic chemicals increased in some of the biospherians' blood after we had been living in Biosphere 2. The reason for this was that we had lost considerable weight in the initial adaptation to our diet, and many of these compounds which are normally tied up in fatty tissue were released into our bloodstreams.

✌ Research with the Russians

There is great interest in Biosphere 2 from the two Russian institutes who have led their country's efforts in developing biological life-support systems for space. The first of these is the Institute of Biomedical Problems in Moscow, where Yevgeny Shepelev became the first human to live with biological life support in 1961 when he spent twenty-four hours in a chamber where chlorella algae regenerated his air and purified his water. At the Institute of Biophysics in Krasnoyarsk, Siberia, these algae-based systems were further developed with the Bios-3 experiments in the 1970s and 1980s, where they achieved six-month closures with a dozen food crops supplying half the food and providing nearly all the air and water regeneration for crews of two and three people. Both institutes have participated in the series of international workshops in closed ecological systems that SBV has organized since 1987 to pull together all the researchers working in this area.

In addition, we do collaborative research with both Russian institutes. Scientists from Siberia worked with us during our pre-closure investigations of the health and medical aspects of closed systems. Their top specialist in microbes essential to human health did a survey of the biospherians to see if their microbial populations would simplify during closure as occurred in the Russian experiments. With the Institute of Biomedical Problems, we had the opportunity to participate in the first space flight study of a functioning, small aquatic ecosystem containing guppies, chlorella, and bacteria. It flew on a Biosatellite and spent twenty-one days in weightlessness in September 1989. In comparison with the closed aquatic systems on the ground set up for a control, the ecosystem functioned perfectly, and all the components of the system reproduced themselves successfully. During our two-year closure, we sent samples of algae growing in varying environments of Biosphere 2 to Moscow for identification. We will work together to evaluate if culturing some of the more productive of these in order to release free oxygen may help to counteract oxygen depletion in Biosphere 2. These results are also applicable to understanding other smaller life-support systems for use in space.

✢✧ Restoration ecology

Biosphere 2 had to be a pioneer in restoration ecology engineering, since all the biomes had to be built from scratch. The reef, for example, had started out as a giant rectangular steel tub. Slowly we filled it with local limestone rock. Then we added salt water made by mixing well water with the Instant Ocean sea salts used in home aquaria. Then we hauled in twenty truck loads of salt water collected from the Pacific Ocean near Scripps Oceanographic Institute in La Jolla, California to provide the basic inoculation of microbiota and plankton. We further seeded the system with algae, reef rock, and invertebrates before introducing the fish and corals. We had recreated a self-sustaining coral reef community that is isolated in a little world, which put us in a unique position to understand what makes reefs tick.

Like rainforests, reefs are endangered around the planet, but we are not sure exactly why. We have some hints: we know that pollution, the dumping of wastes, and the physical impact of boats and divers are all harmful. The overriding question is how to reverse the trend of ocean degradation. How can we restore these damaged areas? What is required to restore the system so that it will resist further human impact and continue to flourish? Restoration ecology is a new field that attempts to answer these questions.

Some restoration ecologists are purists who maintain that the ecosystem must be returned to its original natural form. They feel that not only must the diversity of species and complexity of the community structure be protected, but that man should stay out completely. Given our uncertainty about whether human contact can be made non-destructive, this approach is understandable. At the other end of the spectrum, restoration ecology can be looked at as a science that discovers what the function of an ecosystem is with regard to global or local ecology and then recreates that system to address this need. This is perhaps a more practical approach, which tries to assure that ecosystems are able to evolve and adapt to changing environmental conditions.

Our experience in creating the diversity of ecosystems inside Biosphere 2, and now intensively studying how they develop and mature, will give us powerful tools to apply to restoration ecology in damaged natural systems. Our approach will be one that seeks to unite human intelligence with appropriate technologies that can assist rather than degrade the environment. In the twenty-first century, we can hardly attempt to recreate a world without humans or technology, but we can utilize our strengths to begin to restore what we have degraded.

❧ *A new relationship with the biosphere*

Bill Jordan of the University of Wisconsin Arboretum, a pioneer in restoration ecology, once asked us, "If you can measure anything you want and you know everything about the physical nature of your little world, doesn't it leave you without a feeling for the unknown? Don't you miss going into the wilderness to lose yourself in its beauty and expansiveness? If you can control elements like the rain and temperature, what is there left to seek? Do you experience being responsible for your world as a burden, as a constraint on your freedom?"

Our answer is that quite to the contrary, the experience leaves us with the feeling of hope. For the first time we have a sense of our place in a world where we can see the results of our actions. It is possible to be a positive force in the synergy of life, a necessary element contributing to as well as benefiting from the overall health of the system.

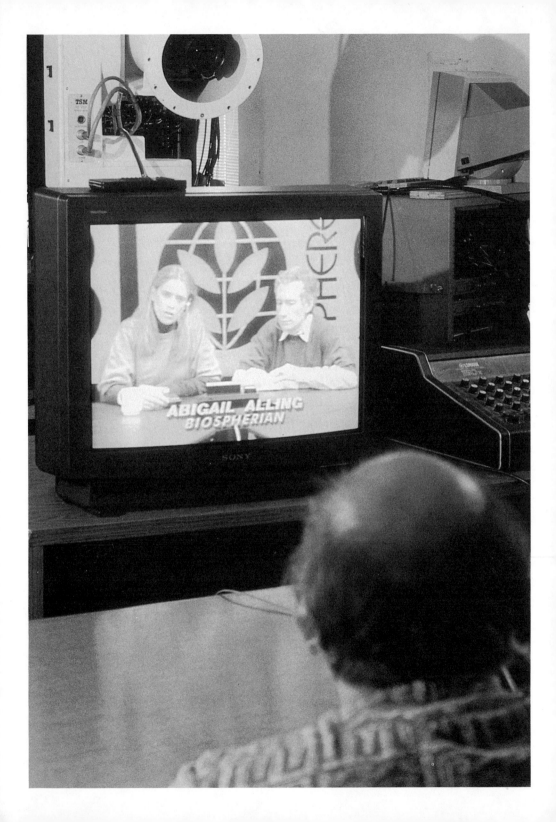

CHAPTER 11

Business Through the Electrons

❧ The paperless office

Offices all over the world are trying to use computers to mini-mize the use of paper, but few are succeeding. In fact, in most offices, the use of computers has probably increased the con-sumption of paper. However, to make a recyclable world, SBV decided on an all-out effort to create a paperless office.

The offices containing our computer equipment are located in the command room, a semi-circular domed room on the second floor of the habitat. Each crew member has a desk with a computer that is equipped with a modem. We have a conference table with the ability to link up with other video sources, and some of the larger computers that can call up all the data from the sensor systems of Biosphere 2 are also housed here. Biosphere 2 has miles of computer, video, and phone cables linking it to Mission Control and the rest of the outside world. So, despite physically leaving the world, we were determined to continue operating in it informationally by learning to use these electronic communications.

The whole staff of Space Biospheres Ventures, including the biospherians, and some of our research consultants in the United States and Europe are connected by an electronic mail system. All of the memos and reports generated inside Biosphere 2 are sent via computer to the outside where they are read on a

We participated in meetings with scientific conferences across the country, as well as with our Mission Control via video conference.

computer screen, printed for distribution, or faxed to the appropriate people. In general, SBV stresses the need to minimize paper use, so employees are encouraged to do as much of their business as possible over the computer lines.

Inside Biosphere 2, we use virtually no paper at all. Since we could neither make it nor re-use it, paper had no place in our world — especially since we would have required tons of it to meet the demands of our communications. It was clear right from the start that our information had to be transmitted and exchanged 'through the electrons'.

The electronic medium makes the most sense both for closed systems like ours and for communication in space exploration. Even in the Earth's biosphere, the production of paper makes heavy demands on the environment. Some of the objectives we had in creating a paperless office were related to how we saw life in the twenty-first century. Our plans exceeded what was available back in 1991. For example, when we came inside in September there was no widely available technology for transmitting faxes without paper. Several months later, the technology to receive faxes on our computer screens came on the market and was incorporated into our systems.

The tempo of our lives, combining sustainable agriculture, naturalist observations, technical operations, and a world-wide network of colleagues, would be impossible without telephones and computers. Biospherians receive mail transmitted via the new electronic fax system. Whereas the various paper-based communications addressed to us are dealt with on the outside by some invaluable assistants who screen our business mail. Unfortunately, junk mail continues to arrive for all of us. Luckily, SBV has a paper recycling system so that this mail can be chucked into recycling bins. Each of us designated someone on the outside to handle personal mail. This is either read to us over a phone, brought to the window, or most commonly, sent in over the electronic fax. If it comes in over the computer screen, we can file it with a label and save it for later reference or reply.

Sometimes, after one of us has done a communications linkup over telephone or video with a school class, thirty or forty letters from the students will come in over our electronic fax system with further questions, their thoughts about becoming biospherians, or even suggestions and recipes for us to try.

Each biospherian was given bound notebooks to use as diaries or journals. They are also used to record scientific data and observations, since some of us still prefer to make our notes by hand in the old-fashioned way. However, with the exception of scientific reference papers, books, and the odd 'post it' pad for note taking, there is virtually no loose paper inside Biosphere 2. When we first started to operate without paper, it was extremely challenging. Now it is a part of our daily life. Some of us installed portable PCs in our apartments. Filing and handling all the hundreds of faxes, reports, memorandums, and other correspondence on a computer system takes far less time than dealing with all those piles of paper that accumulate on a desk.

Just because we lived in a separate world did not mean we got away from the responsibilities of normal citizens on the outside. We could get newspaper articles faxed in, or on the computer, but we could not read the newspaper in the ordinary way. We did watch the news on TV and listen to the radio. Some of the biospherians expressed disappointment that they no longer could see the comics in the newspaper. This became of interest to all eight of us when Biosphere 2 began to be featured in that section. Friends began to tape the comics to a window of the habitat or fax them in for us to see. One of our favorites showed up at Christmas. Santa Claus had landed with his reindeer on the Biosphere 2 spaceframe roof and was shown rapping on the spaceframe structure, trying to get in, with a big question mark hovering over his head!

When we entered the Biosphere, the Soviet Union was one country and apartheid was still legal in South Africa. Nothing else compared in significance to these momentous changes, but there was a good deal of news pertinent to our project. For example, we followed the Environmental Summit in Rio de Janeiro, Brazil with great interest. Especially poignant to us is coverage of environmental disasters such as large oil spills, continued pollution and degradation of the Earth's ecosystems, and the arguments back and forth on who is responsible for this or that environmental crisis. Inside Biosphere 2, we know who is responsible for our world and pointing fingers at wrong-doers is less the issue than taking action to help.

We followed the presidential elections closely. In the fall of 1992, the four of us who are registered voters in Arizona were able to vote in the primary and in the November county, state, and national elections via absentee ballots. Two voting officials from Florence, Arizona (the county seat) came down with the ballots and held them up at the meeting window. Each biospherian voted in turn, reading out over the phone the candidates they wanted to vote for. One of the officials signed on our behalf and then turned in the ballots.

Paying taxes is another responsibility we could not escape. Since we all remained either citizens or residents of the United

States, Uncle Sam required us to pay taxes on our salaries and other income. To make this easier, we each gave someone on the outside power of attorney to sign official documents for us. Tax forms are electronically faxed to us and after we work them out with our accountant or friends, they go out to be signed on our behalf. Then, whomever we have authorized writes a check to the Internal Revenue Service and sends it in. Any refund or salary checks are deposited by our outside helper in our bank accounts. Since our opportunities for spending money are more limited than usual (we get free room and board), many of us have piled up savings like never before. Other mundane matters of life were also handled electronically with little difficulty. These included things like renewing expired driver licenses, making investment and banking decisions, wiring money, and paying bills.

*℣ Managing business at a distance

In the modern world, businesses are increasingly run by people scattered all over the world. We biospherians have also been able to use the communications grid that covers Earth to start new companies, have board of director meetings, and even arrange and participate in scientific workshops and conferences. Why go through all that jet-lag, culture-shock, lost baggage, visa applications, passport and bomb checks, canceled flights, foreign money exchanges, haggling with taxi drivers, hotel reservations, airline food, and other wear and tear on the body and mind when all you want to do is organize and discuss business?

A perfect example is the foundation started by Gaie. Driven by a need to compare the ocean reef in Biosphere 2 to those found in natural areas of the Caribbean where the Biosphere 2 reef was collected, Gaie started a non-profit foundation called the Planetary Coral Reef Foundation (PCRF) shortly before closure. The company addresses questions regarding the demise of reefs world wide, develops techniques to further the restoration ecology of reefs and marshes, and organizes studies of these marine ecosystems in relation to global ecology. The field site chosen was Belize because Blackbird Caye Ltd., a Belizian company, offered to cooperate with PCRF in a joint venture: they would organize an eco-hotel for reef and marsh education and PCRF would set up a research center alongside some of the most beautiful reefs left in the world.

Since the company was only just formed by the time Gaie walked through the airlock door in September 1991, she still had to organize in the legal processes necessary for establishing a non-profit foundation working with Michael McNulty of Brown and Bain P.A., coordinate the team of scientists who would participate in the research program, ensure that income was forthcoming to support the research vessel *Heraclitus*, stationed in Belize, produce a brochure, and conduct meetings with the scientists and board of directors to further the objectives of the foundation. Six months later, she and her colleagues convinced the Belize Government to make the island south of Blackbird Caye a national reserve. PCRF was then leased two acres of land to conduct observations of wild, untouched reefs and marsh lands.

In January of 1992, Laser and a friend initiated another company, Blue Planet Divers, as a subsidiary of PCRF. It was started to provide diving support for the ecotourism business and the foundation's research activities. Laser was also instrumental in establishing a diving school in Belize by purchasing the needed equipment, consulting with the Belize team on business decisions, and encouraging groups, both diving and ecological, to utilize the facilities. All of this was done through computer and phone systems.

Mark was also conducting business over the electrons. He is the chairman and CEO of a small, non-salaried, ecological think tank and research group, the Institute of Ecotechnics. It is registered in England and provides consultation services to a number of innovative ecological ventures around the world, including Biosphere 2. Mark convened and ran two annual meetings of the company and its board of directors while ensconced in front of a computer inside Biosphere 2. Since these meetings of the mind take place through the electronic mail of the twelve directors, there's actually no telling where they may be. He suspects from what he knows of their far-flung activities that they may have participated from computer stations in England, France, Australia, Puerto Rico, Ft. Worth, and Santa Fe, as well as in several offices a stone's throw from his office window. Perhaps one of them is traveling and sending messages during a layover in Singapore or Hong Kong, or while waiting for fog to lift at O'Hare Airport in Chicago.

While this technology is efficient, some may lament that they are unable to read body-language in business meetings held in this fashion. You cannot view the subtle twitching of facial muscle, or the blink of an eye. You are unable to sense eagerness or diffidence, willingness or bluff. On the other hand, conducting business electronically helps to eliminate distractions. Some people are able to communicate better when they don't have a lot of background noise to deal with. Mark was pleased that no one involved in his meetings could hear the noise he was making as he chewed on sugar cane and cracked open roasted peanuts during a break-time snack. Nor could they comment on his messy attire if he was still sporting a couple of mud splats from rice harvesting. All they knew was what appeared on the computer screen.

Roy Walford has continued to oversee the management of the laboratory he runs in UCLA, which employs about ten people. He was able to put together a lengthy grant application to the National Institutes of Health, with portions written at his lab at UCLA and letters of support from the Metabolic Lab in Phoenix and by researchers at the University of Arizona Medical School.

Taber and Jane have even managed to have a house built, which they will move once they leave the Biosphere. They purchased the land, hired the architect, approved designs, and contracted for the building work. While much of this was done through the electrons, it did entail the architect coming by to hold the sketches up to the window for them to see.

❧ Scientific workshops and conferences

Perhaps the most complicated bit of business conducted from inside Biosphere 2 was the arrangement of the Third International Workshop on Closed Ecological Systems hosted by Space Biospheres Ventures and held on the Biosphere 2 site in April 1992. This was a sequel to the first two International Workshops on Closed Systems. The first was sponsored by SBV and the Institute of Ecotechnics; the second was co-sponsored by the Russian Institute of Biophysics. This was, however, the first 'interbiospheric' workshop on closed ecological systems — the participants came from two different biospheres. It was also memorable in that many of the pioneers in the field of closed systems, both from the United States and Russia, participated.

Using the telephone and fax, Mark lined up about thirty scientists from universities and space programs in the U.S., Europe, Russia, and Japan to deliver talks during the four-day event. He even chaired part of the meeting via SBV's two way video system and presented a paper. In addition, all the biospherians were able to participate. During presentations, one video camera took close-ups of slides and graphs and relayed the shots into the Biosphere. Another camera focused on the speaker or questioner from the audience. Besides Mark, Roy and Linda were able to present talks at the workshop by coordinating with someone in the main conference room who presented their slides and graphs.

Five other scientific workshops, on marine systems were organized by Gaie. They involved members of the marine consulting team who were either present on site or convened over the telephone. Sally organized an agriculture workshop and Mark a soils workshop, both of which hosted a number of scientists at Biosphere 2 for meetings spanning several days.

Visiting scientists such as Larry Slobodkin, Richard Evans Schultes, Pat Kangas, and a dozen more gave talks in a guest lecture series which was piped in by video to the biospherians. On other occasions, we ourselves gave scientific lectures and participated in conferences via electronic links. Some of these included talks over a video telephone to the International Space University in Nagoya, Japan; the EcoHarmony conference in Monterey, California; two successive annual black-tie Explorer's Club banquets in New York; and the Case for Mars V Conference in Boulder, Colorado.

The 1992 Annual Explorers Club celebration was a particularly outstanding event. The marine scientist, Sylvia Earle, noted that the meeting was held when explorers now had three worlds to roam, with three distinct life-support systems. Overhead, orbiting in a space shuttle, was her friend Kathy Sullivan, also a member of the Explorers Club. A second world included the 2,000 explorers of planet Earth who had gathered in New York for the meeting. And in the third world were the eight biospherians, explorers of planet Earth's first biospheric offspring.

✽ Educational linkups

Since environmental education is one of the important objectives of Biosphere 2, the crew participates in weekly linkups with schools. These linkups can involve an actual site tour of Biosphere 2 where students interact with biospherians, or they can be conducted through telephone and video contact. Sometimes these link ups involve conferencing many classes, often from several states, over the telephone at the same time. With use of a speaker phone, all the students can ask questions and hear the answers.

The most intriguing linkups have been with classes that designed and built their own 'Biosphere 3' model or mock lunar bases. In one case, thousands of grade-school and high-school students around the U.S. and Canada fabricated a sealed space capsule and lived inside it for three days without coming out. During this time they considered many of the complex aspects of closed systems such as atmosphere regeneration, human health and physiology, communications, command structure and social interactions, agriculture and aquaculture systems, and recycling. Because people learn best by doing, this project is a great tribute to its organizer and all the educators and students who participated in it..

We've also done a couple of wide-reaching educational links via satellite. One was arranged through the respected Fairfax, Virginia school district which helped pioneer 'electronic field trips'. These trips expose students in the U.S. and Canada to exciting research and multicultural encounters. This hookup was seen by about two million students, from grade school to graduate school. On Earth Day 1993, we participated in another major production that was sponsored by IBM. We gave about two and a half million students from Hawaii to New Haven a satellite tour of the Biosphere. Both of these tours featured taped video footage shot in Biosphere 2, live commentary by several crew members, and a question and answer period where students were able to call in directly to the Biosphere and talk with us.

This enthusiastic interest from young people gave us a great boost and enduring hope for the future. Two of the most frequent questions are: "What do I have to do to become a biospherian?" and "What advice would you give us on taking better care of our biosphere?" Though they frequently have a bit of trouble picturing how small Biosphere 2 is — we are often asked about tigers, dolphins, sharks, sometimes even giraffes and whales for their model Biospheres — their enthusiasm confirms our hope that our work will stimulate new ways for people to think about themselves and their environment.

❧ *Electronic campus*

Several of the biospherians have continued their academic careers while inside Biosphere 2. Linda Leigh wrote her application and was accepted to a Ph.D. program with Union Institute of Cincinnati which specializes in innovative graduate programs for multidisciplinary and older students who wish to return to school to complete their doctoral programs. She even took part in her first seminars via phone linkups.

Gaie also prepared her application and was accepted for a Ph.D. program under Dr. Harold Morowitz at George Mason University in Virginia. While she had finished her course work for a doctorate at Yale University before entering Biosphere 2, she had not done her dissertation. She met over Picturetel with her committee from George Mason to discuss her requirements and her proposed thesis to design a biosphere system for a Mars base settlement. Dr. Morowitz was one of the early scientists involved in closed system research as well as being a distinguished thermodynamicist. Fortunately for Gaie, as well as Linda, their acceptance process was simplified by the electrons and neither had to travel across the country for interviews.

Jane and Taber, who had taken advanced technical training with SBV, prepared to enter college programs so they could continue with their interest in space sciences. During our final year in Biosphere 2, they were both accepted into the Honors program at the University of Arizona.

Mark was accepted for a graduate program with the School of Renewable Natural Resources at the University of Arizona. His master's thesis, under faculty advisor Dr. Lloyd Gay, will be a study of an ecological waste recycling system using fast-growing poplar trees. Part of the acceptance requirements involved several prerequisite courses. So during the second year of our closure, Mark signed up for atmospheric and geological science courses. The textbooks came in on one of our monthly import/exports, and Mark was able to complete most of the course assignments on computer files which were then printed out and mailed by our administrative office.

One of Mark's final exams was proctored by Barbara Rossi, a representative from the Extended University of the University of Arizona which arranges these correspondence courses. The test was electronically faxed in, and while Mark worked on answering the questions, Barbara kept an eye on him via the video linkup. At the end, he read off his answers, which she filled in on the paper copy of the exam.

Mark's geosciences final involved drawing diagrams, so the exam was imported for Mark to complete under Roy's supervision. It was then sealed in an envelope and sent out on the next scheduled export. Mark did us proud by 'aceing' all his courses. For the people at Extended University, these correspondence courses with Mark were a new challenge. A few years ago, their most far-out students were soldiers stationed in the Middle East during the Gulf War. Now they had a student in another world!

❧ *The future of electronic participation*

Using our electronic facilities, we've been able to co-author numerous papers for scientific and popular journals. These were sent out electronically and turned into hard copy for editing, study, and review on the outside.

While we have been pleased by how connected we can be

when using the variety of electronic communication devices presently available, we sometimes feel our isolation. Some things that are normally easily arranged take a lot of explaining and haggling to accomplish from a distance — in particular anything that requires a signature. And remember Murphy's Law: what can go wrong will go wrong. Our electronic equipment is fantastic when it works, but a migraine when it doesn't.

Linda gave us a perfect example of how our isolation can cause problems. A State of Arizona wildlife inspector who read in a newspaper story that we had a curved-bill thrasher inside Biosphere 2 visited us to find out if we had the proper licensing papers to keep it in captivity. It took a lot of explaining to make him understand that the bird 'volunteered' for Biosphere 2 by flying in during construction and had consistantly evaded our attempts to export him through the air lock. If filing a paper with him could persuade our feathered friend to willingly depart we'd get someone with our power of attorney to sign in quadruplicate as many forms as he cared to produce! Our problem was that we couldn't catch him, so how could we get him out? Finally, Linda and the inspector worked out all the necessary paperwork to keep everyone happy.

It has been a challenge to leave the world physically and still play an active part in such a wide array of business, scientific, and educational activities. Our flourishing efforts to create a totally paperless office will be improved when some other technologies are added to our system — larger and easier to read computer screens, for example, and a scanner for converting research papers into computer files that can be read, edited, and stored. The challenge of the truly paperless office is still to be met, but we have gone a long way toward doing so.

CHAPTER 12

After Hours

No going out to the movies. No meals in restaurants. No museums to visit or plays to see. No weekends in the mountains. No summer vacations at the seashore. Entertainment was entirely up to the eight of us. We would entertain ourselves or remain un-entertained.

In the early weeks after closure, we had scheduled one-hour crews on Sunday. However, it became apparent quite quickly that we needed time off, time for ourselves and our private pursuits, and that Biosphere 2 could operate itself safely for a day, relying on its alarm system for emergencies. We resorted to a formal vote at a crew meeting in early October to clear the decks of any obligatory meetings on Sundays and to take off all the holidays that the staff on the outside did. The vote was unanimous. If we didn't draw the line, work and research could become grimly obsessive, actually leading to a diminution in the quality of the output. Even so, the necessities of Biosphere 2 meant that someone would be cooking on days off, others feeding animals, and others doing the small chores that couldn't be ignored, like checking water and mechanical systems.

Sunday then became our day of freedom. The only requirement was to show up for breakfast, lunch, and dinner — if you wanted to eat. Otherwise Sunday was our own time to put into

Parties and feasts created special occasions to relax and have fun. Here crew members celebrate Gaie's 34th birthday.

writing, painting, video or music production, rest and relaxa-
tion, or strolling through Biosphere 2 to enjoy a swim or walk
in the rainforest. Before closure, all of our energy went toward
completing the structure and preparing ourselves for the task
inside. Now it had become essential to find leisure time and use
it creatively. We had to make time to play hard as well as work
hard!

Some used Sundays to tour the entire Biosphere and catch
up with the changes that had occurred in areas outside our
specific responsibilities during the past week. This included
snorkeling in the ocean for the three non-divers: Sally, Linda,
and Mark. Four biospherians became avid photographers and
five went on from keeping personal journals and scientific observation notebooks to a variety of other writing projects. Gaie
and Mark used the weekends to work on this book; Mark started
to write poetry; Sally undertook her aforementioned cookbook;
and Laser began an account of his life in Biosphere 2 in his native
tongue, Flemish, for a Belgian publisher. Linda worked on a
naturalist's account of her two years.

The Biospherian Band, led by Jane and Taber, often
jammed on the weekends. We had set up the electronic gear on
the IAB balcony to take advantage of the acoustic effects of the
soaring spaceframe above. Chirping crickets below added an
exotic chorus. Sally played her flute. Jane started to paint. Laser
made a video studio in his room, complete with editing facilities
and an assortment of cameras. He and Gaie worked on interpre-
tive and documentary videos of life inside the Biosphere. Linda
became expert in discoursing on a computer network called the
WELL.

Roy thoroughly enjoyed his special art projects. One of

them involved a montage video documentary which he was putting together. He aimed to create a video art piece that would be displayed on nine separate screens. He also did a photography work where he tried to capture each biospherian in his or her specific field of interest, and he collaborated with Barbara Smith, a performance artist from Los Angeles, on several interactive media events using video phones dealing with Barbara's journey around the planet and Roy's journey inside Biosphere 2.

With the exception of Linda (who didn't want one), we all have TV/VCRs in our apartments. Movies were the most common way to relax. Various members of Mission Control rented movies and then piped them in to us. In addition, seven of us used the TV/cable system. The only source of frustration was our 'video-lagtime.' We'd hear from friends about great new movies and have to wait till they finally were released as videos. Each apartment also has a boombox with AM/FM radio, tape deck, and CD player.

On one occasion we heard that the TV Star Trek program featured an episode about a colony of humans from Earth who were living in a biosphere situated on the polar regions of a barren planet. On Saturday night, we watched a tape of the show on the big screen in the command room with rapt attention. The colonists, who had been genetically engineered to fit into their world of several thousand people, suffered from lack of communication with the outside world. They adhered rigidly to a program formulated two hundred years previously by their founders and were ignorant of current advances in sciences and technology. This regression caused a revolution. Some of their members wanted to re-join humanity by hitching a ride on the starship Enterprise. This portrayed a situation entirely unlike that in our biosphere, where one of our main objectives is to facilitate daily communication flows with the outside world. But the Star Trek show made us contemplate the effect Biosphere 2 was already having on the human imagination.

Since our dinner companions would be the same every night, we tried to vary the agenda and locations. Initially, we made Tuesday night dinners a cultural evening. For a change of scene, we would eat in the Club, the mezzanine room that overlooks the dining room. For the first six months we outlined a series of discussion topics that would rotate between the world of the great explorers (Columbus, Burton, Heyerdahl, Bates, etc.); art (performance art and current art shows in New York); and a series on mythology narrated by Joseph Campbell. Some nights were dedicated to topics relating to other scientists' work that complemented our research inside Biosphere 2.

On other nights we brought up microscopes and explored the intricacies of flowers brought in from the wilderness biomes; or we entered the world of the insect, examining our major adversaries such as the broad mite and thrip; and also our defenders: predatory ladybugs and their larvae which feed on the pests.

Some Tuesdays, we watched topical videos. They ranged from a "History of Insanity," to a documentary on the water and electric utilities of New York City. This last PBS special was fascinating as we compared planet Earth's present 'use and dump' approach to water and sewage with the closed-loop recycling of our new world. The sheer magnitude of sewage that is generated from the people of New York City in one day would fill Yankee Stadium several times over. The sewage is hardly processed at all before it is piped out to the ocean. By contrast, we produced no trash at all; there was no place to put it. Our sewage system was designed to be a carrier of valuable nutrients that we recycle back to our agricultural soil. Although Biosphere 2's rapid buildup time forces its systems to be totally recyclable, even the slower buildup times for the Earth are becoming so full that soon it, too, must use these recycling techniques on a massive scale.

On Thursday nights, Mark Nelson hosted several study series over the two years. We did one on psychology, two on American literature, and several others. Roy initiated a series of autobiographical nights where we told our life stories, one each night. This fascinating weekly event continued through December 1991 and January 1992. Doing this autobiographical review helped to remind us where we had come from and to articulate where we hoped to go.

Friday nights we relocated to an outdoor venue. We had christened the north balcony that overlooks the IAB Café Visionaire and ate dinners at three cafe tables enjoying free-wheeling conversation while the sky transformed from a vivid red sunset to stony black night. While dining al fresco, from time to time we ordered from the attentive French waiter a cappuccino with an eclair or a brandy. Anything could be had at the Café Visionaire — with sufficient imagination.

Saturday nights had no set schedule. But Sunday nights we ate meat, toasted the event with savory fruit juices, and gave speeches or presentations to bring our companions up to speed on any projects we were working on. Not everyone took part; some preferred their privacy. Few organized events interested all eight Biospherians. A diverse crew with diverse tastes, some one or two would nearly always elect not to attend a given event.

But feasts and birthday parties were unanimously appreciated and always guaranteed a full house.

❧ *Parties and feasts*

The first Thanksgiving meal marked the beginning of a tradition of feasts which became an integral part of our culture during the two years. A feast is more than a good meal, it must include superb cuisine, and lots of it, so that the guests can engage in conversation for hours on end while eating to their heart's content. Birthdays gave us a reason to both feast and celebrate. Gaie had the first birthday inside, October 12, with the first Biosphere birthday cake, made with wheat flour, figs, and bananas, and an icing in the shape of an octopus made of frozen papaya and yogurt. Linda's birthday on November 11 featured a great ginger cake with banana yogurt and the first home-made pizza of the closure. By eighteen months later, in March 1993, Jane's and Laser's birthdays became mega-events that included three superb meals for the day highlighted by rationed coffee and home-made ice cream. There was so much food that everyone was able to store some snacks for the following days.

During the second year, choosing a theme and a location for birthday dinners evolved into a favorite game. On Laser's second birthday, we had a campfire picnic on the beach, even though fires are prohibited in Biosphere 2. He had hooked up a TV monitor on the beach where he played a video of a fire burning slowly until it became glowing embers. Some sat around the video screen enjoying our 'fire' while some took a swim in the coral lagoon. Jane's birthday was dedicated to those who work out with weights. We all assembled for the evening meal in the gym room dressed as muscle builders, then later took a sauna with hot steam generated from a little steam pressing machine above which blankets were hung like a teepee. Linda's birthday was centered in her bedroom to tell stories about our favorite dreams. Gaie's birthday was a cybernetic party in the command room where we mounted four video cameras from different angles to monitor the party and played a series of video channels oscillating between Michael Jackson, Madonna, and MTV displayed on the big six-by-eight-foot screen mounted above the table. Taber's July party was in costume; Mark's May birthday was an Australian Outback cowboy affair; Roy's, in June, was an Indian *lungi* party; and Sally's was a formal black-tie dinner.

For Gaie's second birthday inside, Roy had the clever idea of using alcohol from the medical supplies to spike the birthday punch! We all guzzled it down. Now is that desperation — or is it just following the doctor's orders? This lemon/orange ethyl alcohol punch had everyone looped and dancing on tables. It was one of our wildest parties, and we needed it. We have become acutely aware of the importance of good food and stimulants for boosting morale. In a continuing attempt to provide the essential inebriates, Sally tried making *chung* (a Tibetan rice beer) and banana wine, both of which were delicious. But aside from that, there was no alcohol to be had. Good food, plenty of food, and a variety of foods is absolutely essential to extended isolated missions, but let's not forget that stimulants, such as coffee, and inebriants, such as alcohol, loosen everyone up and promote conviviality.

Aside from birthdays, we partied on equinox and solstice days and in general on any holiday we could get away with!

Three-day weekends are important morale-boosters, since we have far too much work to even think about anyone taking a whole week off. In months like August, which lack a biospherian birthday or public holiday, we declared a three-day weekend anyway because it's too long between July 4 and Labor Day!

Solstice and equinox became significant events because they marked the shift of seasons and a change in sunfall patterns. On these days we not only lived it up but contemplated the different routines we must follow with the new season. On our first solstice, December 1991, we called a 'photon feast', glad that we had reached the shortest day of the year. The amount of sunfall would now increase for six months, making our wilderness ecosystems and our agriculture more and more productive. Languidly sprawling on the couches in the plaza, we toasted the sun every time it peaked out from under the clouds.

Sweet potato pies, cheesecakes, trifles, roast pork, a big pot of beans or posole, and coffee if available became the traditional party favorites. Here's the first Thanksgiving menu:

> Baked chicken with water chestnut stuffing
> Sautéed ginger beets
> Baked sweet potatoes
> Stuffed chili peppers sautéed with goat cheese
> Tossed salad with a lemon yogurt dressing
> Orange banana bread
> Tibetan rice beer *(chung)*
> Sweet potato pie topped with yogurt
> Cheesecake
> Coffee with steamed goat's milk

This menu was modest compared to future feasts which ended up with more than three desserts, several types of breads, and sauces or chutneys.

Preparing the feasts was part of the fun, everyone got involved. Sally weighed and measured the whole foods that would be used for each delicacy and generally baked all her treats the day before. Linda, who'd once made a Thanksgiving 'turkey' out of beans for a vegetarian group she lived with, became the number-one vegetable chef. Taber, famous for his patience in spending hours roasting pigs over open fires during pre-closure parties, took on the task of dressing the chickens or pigs, and the rest ducked and danced around one another, chopping, mixing, grinding, and tasting in the crowded kitchen.

How we savored the thought that we would soon have more than enough to eat! No matter how well we adapted to natural foods, we still craved more sweets and stimulants as well as the massive amounts of protein and calories that a normal American diet includes. At the breakfast table, we would fantasize about a spread of eggs, sausage, butter, hash browns, buttered toast, and coffee with cream and sugar instead of porridge, mint tea, and some mashed potatoes. It became a pastime to contemplate where all the food you can buy in supermarkets comes from and where all the trash associated with it ends up. In our little world, this type of opulence and waste was not possible. But on feasts days, feast we did!

To celebrate the anniversary of our first year inside Biosphere 2, Sally baked an enormous cake. Outside, another cake was readied. At 9:30AM that Saturday, Margret Augustine (Biosphere 2's CEO) cut the outside cake just as Sally cut into ours. Theirs was a delicate carrot cake, topped with whipped cream and decorated with a frosting tracery of the Biosphere. Ours was a Mack truck, a heavy, three-story job that Sally could barely manage to hold during the celebratory speech. Mark took the opportunity to weigh his piece — close to a pound and three-quarters. That meant the whole cake weighed fourteen pounds! The giant pieces of cake on our plates, we were told, made quite an impression against our slimmed-down frames. Make a real cake with heavy-duty ingredients like whole wheat flour, ripe bananas, and sweet potatoes, and then top it with thick slices of fig, papaya, and bananas and you have a dish fit for a hungry biospherian. Our friends kidded us about how fast those enormous slices of cake disappeared.

Some of the visitors on that day asked us if we missed quick, convenient food, such as hamburgers or pizza. Well, we do have pizza — it's often the special dish that people request for their birthdays. Delivery is a bit of a problem: our pizza takes four months to make. That's the time it takes to grow a crop of wheat for the flour crust. Then we can add the time for making cheese from our goat's milk and growing the tomatoes, green chilis, and eggplant — without even considering the time and effort necessary to produce any sort of meat topping.

On the Fourth of July, we gathered in the tower library to see fireworks from all sides, the distant ones in Tucson and the Biosphere 2 fireworks out on the grounds. And on New Year's Eve, we climbed the library stairs again to play poker until midnight. Our play was amateurish at best — in the end, Jane bluffed Taber into a large pot, but then lost because she'd gotten confused about whether a straight beats three of a kind or the other way around. None of us would be a threat in high-stakes poker in Las Vegas.

❧ *Arts festival and artists*

A high point of our initial year inside was the first Biospherian Arts Festival in April 1992, a feast for the mind, heart, and stomach. Sally made the first presentation — a wonderful afternoon tea, served on a table beautifully decorated with rainforest flowers. She served demitasse cups of mint tea, along with banana bread, goat's cheese, and banana/papaya jam. While we ate, we made presentations of our works in progress. These included video documentaries, photographs, paintings, poems, and other writings. Jane and Taber teamed up as an electronic band, complete with synthesizer, drums, voice, and pre-recorded 'sounds of the Biosphere'.

Inspired by the unique qualities of our world, Mark began to try his hand at poetry. Encouraged by Roy, who was also writing poetry, Mark hooked up with Roy and some poets at the Electronic Café (a coffee house in Los Angeles) for a joint poetry reading. The contrast between the two frames of reference was striking. Their poems were written in the aftermath of the L.A. riots and reflected the harsh, urgent tempos of the city. Mark's reflected a world where an undercurrent of pumps, sumps, and air handlers underlies a melodic chorus of crickits, the hoots of galagos, the rustling of wind-swept vegetation, and the sounds of a small band of humans who exist in two worlds.

Our second arts festival was in August 1992 and this one was 'interbiospheric'. Jane decorated the interview room till it looked like a discotheque. We invited artists on the SBV staff (Marie Harding who exhibited her recent series of paintings of the Biosphere) and from Tucson (international touring classical guitarist Bill Matthews and artist Peggy Doogan from the University of Arizona) to participate. We set up a video connection to the conference room of the Biosphere 2 Inn and a linkup to the Electronic Café in Los Angeles. Musicians there jammed with Taber and Jane, our Bio Band, Roy and Mark read their poetry, and artists and guests in both worlds shared a wonderful evening.

In August of 1993, with under fifty days left for us in Biosphere 2, we held a second Interbiospheric Arts Festival, which was also a great success. We hope future crews will continue the tradition.

❦ *Strange encounters*

It must have been about 6:00PM on the evening of our first Thanksgiving when over the radio came the oddest sounds of trumpets and horns. We dashed down to the big recreational space on the first floor where we could see a troop of people gathering with flash lights. The management staff of SBV had all gathered, dancing, singing, playing music, and wearing the oddest assortment of clothing. They looked like time travelers visiting from the future. We gazed out into the darkness with astonishment, our breath fogging the windows.

We wondered if we would have Halloween guests come for a trick or treat and sure enough on the first Halloween, some friends did show up. We heard that the most popular Halloween costumes in Tucson that year were groups of eight going around dressed up as biospherians and that one group included a girl with a bandaged finger.

The full moon seemed to really get us going. Everyone seemed wide awake on these nights. It was common to get up for a walk around the Biosphere and meet someone wandering the hallways or paths, or even meet night-owl friends roaming the outside of the Biosphere peering in!

One of the most interesting (and distant) links we made was in March 1992 with the twenty-two-person American research team at the South Pole. Part of our fascination was in finding a mirror through which to see ourselves. Were we like them? What kind of explorers were we? The connection was achieved via phone from the Biosphere to a ham radio operator in South Dakota who was connected with members of the crew packed into their radio shack.

They had been there over five months, and as we spoke, it was the eve of their entry into the six months of night when the sun dips below the horizon. They told us it was minus seventy degrees Fahrenheit outside with a wind chill factor of minus one hundred forty. We regaled them with tales of our lush tropical world in Arizona. Planes can't land for eight months out of twelve in Antarctica, so the videos and books on Biosphere 2 we sent them had to await their June mid-winter mail drop. We both live in domes, but for obvious reasons theirs aren't made of glass. Evidently their main building, a fifty-foot-diameter geodesic dome, is on shifting and settling ice, so they're constantly chocking it up to stabilize it. They have a station cook who works six days a week; they forage for themselves on Sundays.

Their individual quarters are tiny — six-by-ten-foot rooms, which house two crew members during the summer months. The doctor and station head have bigger quarters because their rooms are also used for medical facilities and emergencies. At the time we spoke, they were just setting up an eight-by-ten-foot space as a greenhouse, using hydroponics and high sodium lights to grow winter vegetables. They also do home-brewing like we do, although they have a good supply of beer. We survived quite well without beer, but we weren't sure we could handle six months without the sun!

✤ *Families and friends*

You can prepare for saying good-bye, but in some ways no preparation is enough. Some biospherians wondered what would happen when they physically separated from their partners for two years. And what about families? It took some getting used to, but we made our family lives thrive by means of window meetings, telephone, e-mail, electronic faxes, two-way radio, and video technology. Picturetel, an interactive video system we have inside, relays real-time images and sound from any two parties that have the same system, just like a satellite linkup, but it's done through special phone lines. The first Christmas in the Biosphere, all the biospherians were linked up to their families for a one-hour meeting, whether they were in London, Antwerp, California, or New York.

Mark made a Chanukah linkup with his mother, Minnie, in New York who had his aunt and two friends there to chat with her son. These are women in their seventies and eighties who had come to America from the Old World and were visiting by way of a futuristic high-tech medium. Mark enthralled them with a nine-minute video Laser had made of life and work inside Biosphere 2. What struck them especially was the contrast between the labor of growing our own crops and our high-tech kitchen. They enthusiastically suggested some traditional Jewish recipes to add to our repertoire. Minnie sent him a recipe for a sweet potato pie (kugel) and his aunt sent one for gefilte fish which we could make using our tilapia fish. In the long run, our contacts with our families and friends became the magic ingredient that made Biosphere 2 a place to call home.

The Adventure Never Ends

❧ Changes in a new world

After eighteen months we all agreed that we really felt at home, content in our adaptation to this new world. It wasn't easy, but we had finally adjusted to a new diet, a new work schedule, a different social milieu, a dramatically different atmosphere with falling oxygen and fluctuating carbon dioxide, and the unique seasonal patterns of our biomes. And just then, it was time to prepare to leave and hand over our activities to a new crew.

At the second New Year's Day we celebrated in Biosphere 2, Laser commented that sometimes his body seemed to adapt and other times it seemed to fall out of synch. Dr. Oleg Gazenko, a pioneer in Russian space medicine, had succinctly described the acid test of successful adaptation to new conditions when he visited Biosphere 2: "You will know your body has adapted when you feel a sense of freedom." By March of 1993 many of us had independently experienced a distinct change. There was a deepened feeling of unity among the eight, a feeling of rest and well being. Eighteen months had been a long haul, with many ups and downs, but with the realization that we would now surely complete the mission, it indeed was a feeling of freedom.

At the end of its two-year maiden voyage, Biosphere 2 is like a well-tested ship, ready to carry on research in biospherics for decades to come.

In checking the literature on small groups in isolated environments, Roy had found that frequently the third quarter was the most difficult. The fourth and final quarter is usually far easier, with high morale returning in anticipation of completion. This matched our experience in Biosphere 2. Our third quarter had been the one with the severest oxygen depletion, the dark and stormy winter with its impact on both food crops and our good spirits, and the greatest disharmony among the crew. By contrast, our last six months were marked by sunny skies, high morale, a deepened sense of physical well-being, and an emotional harmony that could ride the ups and downs.

It took the full two years to adjust. Future biospherians may find it easier, since we spent much of the first year debugging the system, finding out what would go wrong or be the greatest challenges. Our second year shifted to fine-tuning systems and operations and expanding research endeavors. And of course we also had more time to relax and absorb the experience. It was a time for discovery and contemplation of our role in this new world. We had gained the detachment to appreciate the enormous effort which the two years had demanded. It seemed as though we had touched every aspect of our world; we interacted with molecules and with trees, we knew our environment's boundaries and its subtleties.

We anticipated our departure with nostalgic sadness. There was no doubt in our minds that at 8:15AM on September 26, 1993 we would struggle with the departure. It would be a sharp transition, even though only a door kept the two worlds apart. It was as if we had departed the planet for two years and in some sense we surely had.

❧ *Handovers and upgrades*

It was in our last six months that we were forced to get deeply involved with the planning for Transition One. This is the anticipated transition phase between this crew and the next. Gaie spent many hours a day working with John Allen, Vice-President of Biospheric Development, and Jack Corliss, the Director of Research, on the organization of the transition research and development. Sally worked with Norberto Alvarez-Romo, Vice President of Mission Control, to develop the biospherian training program and selection process for future teams, while organizing the process by which the crew would hand over operations to the new biospherians. Laser and Larry Pomatto, Director of Technical Operations, organized the technical changes that would be required during the transition.

The transition would be a challenging time because, among many factors, we aimed to maintain the integrity of Biosphere 2's atmosphere. That means that people are permitted to enter the Biosphere using the airlock system so long as they do not exceed 192 population-hours a day. This number of hours is equivalent to having eight biospherians live in the biosphere for twenty-four hours. The amount of carbon dioxide released and oxygen consumed will be the same. This procedure ensures that our long-term monitoring of atmospheric cycles can continue practically undisturbed. Because there will be considerable changes made in technical systems, this transition will take five months. We anticipate that its final two months will focus on the next crew and their final training before they begin their mission. We think that future transitions between missions can be reduced to about six weeks for research and operation handovers.

In taking Biosphere 2 from theory to practice, we gained clear ideas of what worked and what didn't. The few technical systems that will be replaced almost all involve simplifying the original system. Computer-controlled systems have a tendency to become more complex than necessary. In Biosphere 2, we had to learn to simplify the design so it only accomplishes what is necessary.

Sally was in charge of teaching the new team the operating techniques for the entire system. Although the Mission Two team had not at that point been finally selected, the pool of applicants started their training on site during our second year of closure. It's an extraordinarily culturally diverse set of men and women. Candidates again range from twenty-year-olds to those in their sixties and are from many nations, including the former Yugoslavia, Australia, Mexico, Nepal, Germany, Britain, and the United States. Choosing the right mix of people is vital because they must have skills diverse enough to handle the hundreds of systems, the knowledge to carry out a wide variety of research projects, and the ability to interact with the range of people interested in Biosphere 2, from scientists, to environmentalists, to elementary school children.

During the tight time-table for people allowed into the Biosphere during transition, research scientists take a high priority to make complete surveys of insects, animals, plants, soils, and water and to introduce new species. Every plant in the wilderness biomes will again be tagged and measured, which will enable us to record growth and losses. In the ocean, we will not only make some technical upgrades, but we will also enhance the food web to include more herbivores (fish, sea urchins, and snails) to handle the proliferation of plants (micro- and macro-algae). All the colonies of corals being tracked will be tagged and mapped to follow them through time. We will add roughly twenty new woody trees each to the desert, savannah, and marsh.

In the rainforest, a major objective will be to increase the number and variety of edible species. In particular we will be adding more productive types of bananas, taros, papayas, and other distinctive rainforest food-plants. In the cloud forest we will attempt to match a new plant mix to the environmental conditions there. With the planned removal of our initial tree canopy species, we can also widen the biodiversity of the forest floor and lower canopies of the lowland rainforest. In the agriculture biome, we will enrich the diversity of low-light and understory crops and include more supplemental artificial lighting in shady areas.

We will no longer be permanent residents of the Biosphere, but we will remain quite involved in it as we enter and exit through the airlock to transfer our know-how to others. When we move out on September 26th, our live-in crew of eight will be replaced by a single, rotating, nighttime watch member during the transition phase until the crew of Mission Two moves in. During the day, Mission Two crew will be doing the agriculture and basic operations about four hours per day.

❧ *The impossible handover*

We can hand over the controls, but it will be almost impossible to hand over our experience. We took for granted the cycles of water or the composition of the air we breathe — until we faced our daily role in their maintenance. This is a totally new experience, a way of life we discovered by living inside an artificial biosphere. Out there in Biosphere 1 every action also has its consequences, even if we have difficulty understanding them.

It was the founder of biospheric theory, a Russian geochemist named Vladimir Vernadsky, who was among the first to see, in the early 1900s, that humanity was altering Earth's biosphere with the expansion of cities, industry, technology, travel, agricultural production, and pollution. Man, he said, is a powerful geological force. Vernadsky had the vision to observe, long before the environmental crises of our time, that we would need to evolve in order to reconcile the conflict between technology and ecology. Man would have to become a more intelligent manager of the biosphere or face the threat of irreparably harming its ability to sustain life.

Man's invention of tools and technologies is an extension of life's inherent drive to continually extend its domain. Our invention of spaceships coupled with our ability to create biospheres may enable life on Earth to expand into the solar system and beyond in the millennia to come. This is exactly what Vernadsky had suggested in his far-reaching conclusions. Biospheres are a cosmic phenomenon, linked to the universe through our dependence on energy from the sun. But far from being confined to any originating planet, biospheres in the fullness of time will spread life throughout space. Biosphere 2 is undergoing an evolutionary process, one partially molded by those who live inside it, but the people are a part of that evolution as well. Man's challenge is to learn to evolve in harmony with the biosphere instead of in conflict with it. This is not easy — it may require profound shifts in our thinking and in our ways of acting.

❧ *Leaving*

Although we chatted from time to time about leaving, none of us really grasped its impact. How could we anticipate what the shock of transition would be like? Physiologically and psychologically there would be major adjustments, just as there had been when we entered. We hadn't known what to expect then, either.

In some ways, we have been renewed by our experience. Our cholesterol levels are still at extraordinarily low levels. Our blood pressure has lowered and our pulses have slowed. Complexions are clear. We are exposed to almost no ultraviolet radiation, and we are eating extraordinarily fresh and uncontaminated foods, breathing biologically cleansed air, and drinking pure water. Yet mastering existence inside Biosphere 2 had been a struggle, and every step of the way was hard-won. So who are we, now that we approach re-entry, and do we even fully realize the changes we have undergone?

Mark made a speech during his second birthday celebrating inside Biosphere 2. He concluded with an acknowledgment that he was quite simply happy. It was amusing to hear our erstwhile number one grouch declaring life to be happy and good. Even after all the sweat and tears, and not a few heated discussions, we have become a united team dedicated to the task of keeping our little world in fine shape.

✣ *Food fantasies and family suggestions*

Part of the preparation to leave involved deciding where we were going to live. Some, feeling the desire to travel again in the world, put some of their energy into getting cars so they would have wheels the first day out! There was a great deal of discussion about buying electric cars to help reduce fuel emissions, but in reality there is still no such car on the market capable of covering long distances. Sally shopped for a black evening gown to wear during the first dinner which was to be held on the green lawn just outside the habitat on the night of re-entry. Her friend brought the latest Spiegel catalogue to the window so she could order her dress in advance!

All of us dreamed about which foods we would first devour. Taber announced one morning at breakfast that he was going to ask his father to bring a giant chest of food so that during the first day anyone could discreetly disappear for a minute, delve into the chest, and grab some goodies. Carol Hemingway, a media consultant from Los Angeles, promised to bring Laser a snickers-bar cheesecake (he had heard it was a California specialty). Gaie said she wanted champagne — a case would do — and Sally's mouth watered for lox and bagels. Mark dreamt of a breakfast of freshly baked croissants and a cappuccino. Roy, who was forever trying to entice someone to take pity on the biospherians and convince Mission Control to allow in a case of Scotch, wanted a bottle of his favorite brand waiting for him when he stepped out the door. In a departure from the food fixations of the rest of us, Linda most looked forward to being with the people dearest to her and a trip into the wide, wild open spaces.

Many of us felt a concern, or perhaps just curiosity, about what it would be like to interact with all the people outside! There is the luxury of great privacy inside the spaciousness of Biosphere 2, where it was rare to pass others in the hallway, and we had to carry two-way radios to stay in touch. But on the outside, we would be constantly around people who say hello and want a response! There would be major adjustments to make in the fast-paced, people-filled world of Biosphere 1. We looked forward to ending our isolation, but at the same time we realized with a bit of trepidation that we would no longer have the walls to protect us.

❧ *Personal preparation*

We had anticipated that Biosphere 2 would be as much a time machine as a journey in finite space. The ceaseless transformations and movements of matter are quickened since life and technology are so concentrated. Time seems to have expanded at the same time our space has condensed. For example, since our air and water cycles seem to be hundreds if not thousands of times faster than the equivalent Earth cycles, one Biosphere 2 year is a far greater time-event than an Earth year. Perhaps this is one reason why the experience in Biosphere 2 is so rich and deep.

Rusty Schweickart, a crew member on Apollo 9, served on the original Project Review Committee of Biosphere 2. Rusty has a unique ability to eloquently describe his experience watching Earth from space as an orbiting globe, a unified whole without the national divisions that politics and mapmakers impose. At the first Biosphere 2 workshop in December 1984, he spoke about his space experience with an enlightening perspective.

"You are up there as the sensing element, that point out on the end, and that's a humbling feeling. It's a feeling that says you have a responsibility. It's not for yourself.... And when you come back there's a difference in the world now. There's a difference in the relationship between you and that planet and you and all those other forms of life on that planet....It's a difference and it's so precious. And all through this I've used the word 'you' because it's not me, it's not Dave Scott, it's not Dick Gordon, Pete Conrad, John Glenn — it's you, it's we. It's Life that's had that experience."

When someone asked Rusty how he had prepared to go to space, he answered that he prepared for his momentous task by reviewing the great thinkers of the world, the poets and philosophers who spoke most clearly to his heart. Joe Allen, the shuttle astronaut, whose Midwestern, Will-Rogers sense of humor and drawl disguise the fact that he holds a Ph.D. in physics from Yale, had quite a different perspective. When asked the same question, he replied, "The night before my flight I studied the operating manuals."

Like Rusty, we felt that we were granted the privilege to experience a new relationship to our global biosphere by inhabiting a man-made one. Like Joe, we know that without operating manuals we cannot do our job. Perhaps these are the most important approaches that we can use to share our experience with others.

In his analysis of the core mythology that lies behind all human culture, Joseph Campbell traced what he calls "the hero's path." The first stage is the vision, or calling. Then comes the adventure in a new world, the land of magic and mystery, where many obstacles must be overcome but great wisdom can be won by the survivor. The last, and perhaps most difficult stage, is the return. The hero must re-enter the ordinary world and help to change it in accordance with the insight he has gained on his vision quest. While we do not view ourselves as heroes, this sequence has great resonance for us. Our greatest challenge is yet to come — to communicate the reality we experienced living under glass.

Index